# Global Navigation Satellite Systems and their Applications

# Global Navigation Satellite Systems and their Applications

Edited by
**Asher Clark**

▤ Larsen & Keller
www.larsen-keller.com

Global Navigation Satellite Systems and their Applications
Edited by Asher Clark
ISBN: 978-1-63549-137-1 (Hardback)

© 2017 Larsen & Keller

# ▤ Larsen & Keller

Published by Larsen and Keller Education,
5 Penn Plaza,
19th Floor,
New York, NY 10001, USA

**Cataloging-in-Publication Data**

Global navigation satellite systems and their applications / edited by Asher Clark.
    p. cm.
Includes bibliographical references and index.
ISBN 978-1-63549-137-1
1. Global Positioning System. 2. Artificial satellites in navigation. 3. Satellites. I. Clark, Asher.
G109.5 .G56 2017
910.285--dc23

The publisher's policy is to use permanent paper from mills that operate a sustainable forestry policy. Furthermore, the publisher ensures that the text paper and cover boards used have met acceptable environmental accreditation standards.

Printed and bound in the United States of America.

For more information regarding Larsen and Keller Education and its products, please visit the publisher's website www.larsen-keller.com

# Table of Contents

# Preface

This book is a compilation of chapters that discuss the most vital concepts in the field of satellite navigation or global navigation satellite systems. It explores all the important aspects of this field in the present day scenario. Satnav is the technology that uses satellites to provide particular locations by using time signals emitted by radio. It is applied in Global Positioning Systems (GPS) worldwide and is being used by many different countries for their military and security purposes. While understanding the long-term perspectives of the topics, the textbook makes an effort in highlighting their impact as a modern tool for the growth of the discipline. It presents the complex subject of satellite navigation in the most comprehensible and easy to understand language. This textbook is appropriate for graduates and post-graduates in the fields of maritime communication, wireless and digital technologies and avionics.

A detailed account of the significant topics covered in this book is provided below:

Chapter 1- Global navigation satellite systems are used to provide geo-spatial data and information to users by using satellites. These systems function by electronic transmitters that provide precise location using time signals. This chapter serves as an introduction to global navigation satellite systems.

Chapter 2- Satellite navigation systems are required to be accurate to provide real-time information. Reliable augmentation of global navigation satellite systems has been made possible through satellite-based augmentation systems, ground-based augmentation and additional navigational sensors like automated celestial navigation systems, inertial navigation systems etc. This chapter provides comprehensive information on the classification of satellite navigation systems.

Chapter 3- There are several satellite navigation systems in use globally. This chapter provides the reader with elaborate material on different satellite navigation systems such as GPS Aided GEO Augmented Navigation, TRANSIT system, BeiDou Navigation Satellite System and Galileo. The content describes the main objectives, international involvement, system description and impact of these systems on modern technology.

Chapter 4- Global positioning system (GPS) is a global satellite navigation system that provides the user with precise coordinates of locations present anywhere on the earth. This chapter studies the global positioning system thoroughly by exploring its components like GPS satellite blocks, GPS signals, GPS navigation device, GNSS positioning calculation, GNSS enhancement and error analysis for the global positioning system. All the main components of global positioning system are explained here.

Chapter 5- Satellite navigation systems are used in a variety of ways and this chapter focuses on vehicle tracking systems and automated vehicle locations. The content delves into the types of systems used in these applications and the conventional and unconventional uses of each. A section also talks about an important aspect of GPS systems namely, dilution of precision and its components.

Chapter 6- All navigation systems are dependent on satellites orbiting the atmosphere for providing accurate information. A satellite is a man-made object that is launched into the orbit of the Earth by a spacecraft. This chapter illustrates the major satellite subsystems such as satellite bus, fractioned spacecraft, telemetry, GNSS software-defined receiver and spacecraft propulsion. The chapter serves as a source to understand the major categories related to satellites.

Chapter 7- Satellites can be categorized by their various applications in space telescopes, communications satellite, Earth observation satellites, weather satellites, reconnaissance satellites and biosatellites. The chapter provides an in-depth analysis of each type with suitable examples to enable better understanding.

Chapter 8-Geodesy or geodetic engineering is the three-dimensional representation of the Earth's surface by surveying specific regions. It helps in positioning within the temporally varying gravitational field. It is also vital in cartography. This section gives a comprehensive understanding of geodesy to the readers.

Chapter 9- This chapter inspects the various satellites used in positioning based on region. The satellites studied include Indian Regional Navigation Satellite System (India), Quasi-Zenith Satellite System (Japan) and DORIS (France). The reader is provided with information on the orbit, positioning and special functions. The aspects elucidated in this chapter are of vital importance, and provide a better understanding of satellite navigation systems.

I would like to make a special mention of my publisher who considered me worthy of this opportunity and also supported me throughout the process. I would also like to thank the editing team at the back-end who extended their help whenever required.

**Editor**

# Introduction to Global Navigation Satellite Systems

Global navigation satellite systems are used to provide geo-spatial data and information to users by using satellites. These systems function by electronic transmitters that provide precise location using time signals. This chapter serves as an introduction to global navigation satellite systems.

## Satellite Navigation

A satellite navigation or satnav system is a system that uses satellites to provide autonomous geo-spatial positioning. It allows small electronic receivers to determine their location (longitude, latitude, and altitude/elevation) to high precision (within a few metres) using time signals transmitted along a line of sight by radio from satellites. The system can be used for navigation or for tracking the position of something fitted with a receiver (satellite tracking). The signals also allow the electronic receiver to calculate the current local time to high precision, which allows time synchronisation. Satnav systems operate independently of any telephonic or internet reception, though these technologies can enhance the usefulness of the positioning information generated.

A satellite navigation system with global coverage may be termed a global navigation satellite system (GNSS). As of April 2013 only the United States NAVSTAR Global Positioning System (GPS) and the Russian GLONASS are global operational GNSSs. China is in the process of expanding its regional BeiDou Navigation Satellite System into the global Compass navigation system by 2020. The European Union's Galileo is a global GNSS in initial deployment phase, scheduled to be fully operational by 2020 at the earliest. India currently has satellite-based augmentation system, GPS Aided GEO Augmented Navigation (GAGAN), which enhances the accuracy of NAVSTAR GPS and GLONASS positions. India has already launched the IRNSS, with an operational name NAVIC (Navigation with Indian Constellation), a constellation of satellites for navigation in and around the Indian Subcontinent. It is expected to be fully operational by June 2016. France and Japan are in the process of developing regional navigation systems as well.

Global coverage for each system is generally achieved by a satellite constellation of 20–30 medium Earth orbit (MEO) satellites spread between several orbital planes. The actual systems vary, but use orbital inclinations of >50° and orbital periods of roughly twelve hours (at an altitude of about 20,000 kilometres or 12,000 miles).

## Classification

Satellite navigation systems that provide enhanced accuracy and integrity monitoring usable for civil navigation are classified as follows:

- GNSS-1 is the first generation system and is the combination of existing satellite navigation systems (GPS and GLONASS), with Satellite Based Augmentation Systems (SBAS) or Ground Based Augmentation Systems (GBAS). In the United States, the satellite based component is the Wide Area Augmentation System (WAAS), in Europe it is the European Geostationary Navigation Overlay Service (EGNOS), and in Japan it is the Multi-Functional Satellite Augmentation System (MSAS). Ground based augmentation is provided by systems like the Local Area Augmentation System (LAAS).

- GNSS-2 is the second generation of systems that independently provides a full civilian satellite navigation system, exemplified by the European Galileo positioning system. These systems will provide the accuracy and integrity monitoring necessary for civil navigation; including aircraft. This system consists of L1 and L2 frequencies (in the L band of the radio spectrum) for civil use and L5 for system integrity. Development is also in progress to provide GPS with civil use L2 and L5 frequencies, making it a GNSS-2 system.[1]

- Core Satellite navigation systems, currently GPS (United States), GLONASS (Russian Federation), Galileo (European Union) and Compass (China).

- Global Satellite Based Augmentation Systems (SBAS) such as Omnistar and StarFire.

- Regional SBAS including WAAS (US), EGNOS (EU), MSAS (Japan) and GAGAN (India).

- Regional Satellite Navigation Systems such as China's Beidou, India's NAVIC, and Japan's proposed QZSS.

- Continental scale Ground Based Augmentation Systems (GBAS) for example the Australian GRAS and the US Department of Transportation National Differential GPS (DGPS) service.

- Regional scale GBAS such as CORS networks.

- Local GBAS typified by a single GPS reference station operating Real Time Kinematic (RTK) corrections.

## History and Theory

The first satellite navigation system was Transit, a system deployed by the US military in the 1960s. Transit's operation was based on the Doppler effect: the satellites travelled on well-known paths and broadcast their signals on a well-known frequency. The received frequency will differ slightly from the broadcast frequency because of the movement of the satellite with respect to the receiver. By monitoring this frequency shift over a short time interval, the receiver can determine its location to one side or the other of the satellite, and several such measurements combined with a precise knowledge of the satellite's orbit can fix a particular position.

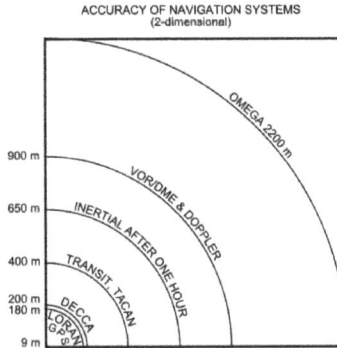

Early predecessors were the ground based DECCA, LORAN, GEE and Omega radio navigation systems, which used terrestrial longwave radio transmitters instead of satellites. These positioning systems broadcast a radio pulse from a known "master" location, followed by a pulse repeated from a number of "slave" stations. The delay between the reception of the master signal and the slave signals allowed the receiver to deduce the distance to each of the slaves, providing a fix.

Part of an orbiting satellite's broadcast included its precise orbital data. In order to ensure accuracy, the US Naval Observatory (USNO) continuously observed the precise orbits of these satellites. As a satellite's orbit deviated, the USNO would send the updated information to the satellite. Subsequent broadcasts from an updated satellite would contain the most recent accurate information about its orbit.

Modern systems are more direct. The satellite broadcasts a signal that contains orbital data (from which the position of the satellite can be calculated) and the precise time the signal was transmitted. The orbital data is transmitted in a data message that is superimposed on a code that serves as a timing reference. The satellite uses an atomic clock to maintain synchronization of all the satellites in the constellation. The receiver compares the time of broadcast encoded in the transmission of three (at sea level) or four different satellites, thereby measuring the time-of-flight to each satellite. Several such measurements can be made at the same time to different satellites, allowing a continual fix to be generated in real time using an adapted version of trilateration.

Each distance measurement, regardless of the system being used, places the receiver on a spherical shell at the measured distance from the broadcaster. By taking several such measurements and then looking for a point where they meet, a fix is generated. However, in the case of fast-moving receivers, the position of the signal moves as signals are received from several satellites. In addition, the radio signals slow slightly as they pass through the ionosphere, and this slowing varies with the receiver's angle to the satellite, because that changes the distance through the ionosphere. The basic computation thus attempts to find the shortest directed line tangent to four oblate spherical shells centred on four satellites. Satellite navigation receivers reduce errors by using combinations of signals from multiple satellites and multiple correlators, and then using techniques such as Kalman filtering to combine the noisy, partial, and constantly changing data into a single estimate for position, time, and velocity.

## Civil and Military Uses

Satellite navigation using a laptop and a GPS receiver

The original motivation for satellite navigation was for military applications. Satellite navigation allows the precision in the delivery of weapons to targets, greatly increasing their lethality whilst reducing inadvertent casualties from misdirected weapons. Satellite navigation also allows forces to be directed and to locate themselves more easily, reducing the fog of war.

The ability to supply satellite navigation signals is also the ability to deny their availability. The operator of a satellite navigation system potentially has the ability to degrade or eliminate satellite navigation services over any territory it desires.

## Global Satellite Navigation Systems

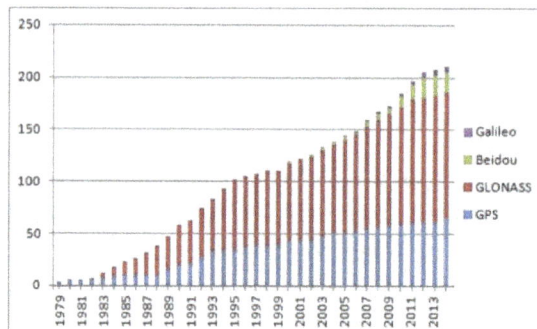

launched GNSS satellites 1978 to 2014

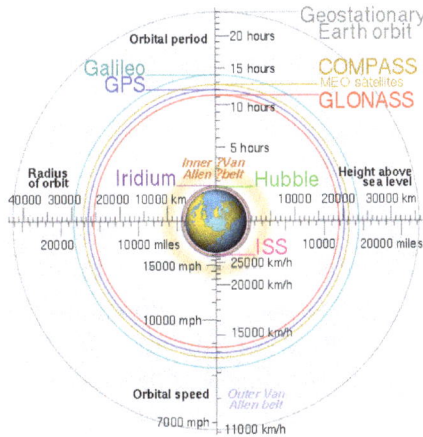

Comparison of geostationary, GPS, GLONASS, Galileo, Compass (MEO), International Space Station, Hubble Space Telescope and Iridium constellation orbits, with the Van Allen radiation belts and the Earth to scale.The Moon's orbit is around 9 times larger than geostationary orbit. (In the SVG file, hover over an orbit or its label to highlight it.)

## Operational

## GPS

The United States' Global Positioning System (GPS) consists of up to 32 medium Earth orbit satellites in six different orbital planes, with the exact number of satellites varying as older satellites are retired and replaced. Operational since 1978 and globally available since 1994, GPS is currently the world's most utilized satellite navigation system.

## GLONASS

The formerly Soviet, and now Russian, *Global'naya Navigatsionnaya Sputnikovaya Sistema* (Russian: ГЛОбальная НАвигационная Спутниковая Система, GLObal NAvigation Satellite System), or GLONASS, is a space-based satellite navigation system that provides a civilian radionavigation-satellite service and is also used by the Russian Aerospace Defence Forces. The full orbital constellation of 24 GLONASS satellites enables full global coverage.

## In Development

## Galileo

The European Union and European Space Agency agreed in March 2002 to introduce their own alternative to GPS, called the Galileo positioning system. At an estimated cost of EUR 3.0 billion, the system of 30 MEO satellites was originally scheduled to be operational in 2010. The original year to become operational was 2014. The first experimental satellite was launched on 28 December 2005. Galileo is expected to be compatible with the modernized GPS system. The receivers will be able to combine the signals from both Galileo and GPS satellites to greatly increase the accuracy. Galileo is now not expected to be in full service until 2020 at the earliest and at a substantially higher cost. The main modulation used in Galileo Open Service signal is the Composite Binary Offset Carrier (CBOC) modulation.

## BeiDou

China has indicated they plan to complete the entire second generation Beidou Navigation Satellite System (BDS or BeiDou-2, formerly known as COMPASS), by expanding current regional (Asia-Pacific) service into global coverage by 2020. The BeiDou-2 system is proposed to consist of 30 MEO satellites and five geostationary satellites. A 16-satellite regional version (covering Asia and Pacific area) was completed by December 2012.

## Regional Satellite Navigation Systems
### BeiDou-1

Chinese regional (Asia-Pacific, 16 satellites) network to be expanded into the whole global system which consists of all 35 satellites by 2020.

## NAVIC

The NAVIC or NAVigation with Indian Constellation is an autonomous regional satellite navigation system developed by Indian Space Research Organisation (ISRO) which would be under the total control of Indian government. The government approved the project in May 2006, with the intention of the system completed and implemented on 28 April 2016. It will consist of a constellation of 7 navigational satellites. 3 of the satellites will be placed in the Geostationary orbit (GEO) and the remaining 4 in the Geosynchronous orbit(GSO) to have a larger signal footprint and lower number of satellites to map the region. It is intended to provide an all-weather absolute position accuracy of better than 7.6 meters throughout India and within a region extending approximately 1,500 km around it. A goal of complete Indian control has been stated, with the space segment, ground segment and user receivers all being built in India. All seven satellites, IRNSS-1A, IRNSS-1B, IRNSS-1C, IRNSS-1D, IRNSS-1E, IRNSS-1F, and IRNSS-1G, of the proposed constellation were precisely launched on 1 July 2013, 4 April 2014, 16 October 2014, 28 March 2015, 20 January 2016, 10 March 2016 and 28 April 2016 respectively from Satish Dhawan Space Centre. The system is expected to be fully operational by August 2016.

## QZSS

The Quasi-Zenith Satellite System (QZSS), is a proposed three-satellite regional time transfer system and enhancement for GPS covering Japan. The first demonstration satellite was launched in September 2010.

## Augmentation

GNSS augmentation is a method of improving a navigation system's attributes, such as accuracy, reliability, and availability, through the integration of external information into the calculation process, for example, the Wide Area Augmentation System, the European Geostationary Navigation Overlay Service, the Multi-functional Satellite Augmentation System, Differential GPS, and Inertial Navigation Systems.

## DORIS

Doppler Orbitography and Radio-positioning Integrated by Satellite (DORIS) is a French precision navigation system. Unlike other GNSS systems, it is based on static emitting stations around the world, the receivers being on satellites, in order to precisely determine their orbital position The system may be used also for mobile receivers on land with more limited usage and coverage. Used with traditional GNSS systems, it pushes the accuracy of positions to centimetric precision (and to millimetric precision for altimetric application and also allows monitoring very tiny seasonal changes of Earth rotation and deformations), in order to build a much more precise geodesic reference system.

## Low Earth Orbit Satellite Phone Networks

The two current operational low Earth orbit satellite phone networks are able to track transceiver units with accuracy of a few kilometers using doppler shift calculations from the satellite. The coordinates are sent back to the transceiver unit where they can be read using AT commands or a graphical user interface. This can also be used by the gateway to enforce restrictions on geographically bound calling plans.

## References

- "Commission awards major contracts to make Galileo operational early 2014". 2010-01-07. Retrieved 2010-04-19.

- "India to build a constellation of 7 navigation satellites by 2012". Livemint.com. 2007-09-05. Retrieved 2011-12-30.

- S. Anandan (2010-04-10). "Launch of first satellite for Indian Regional Navigation Satellite system next year". Beta.thehindu.com. Retrieved 2011-12-30.

# 2

# Classification of Satellite Navigation Systems

Satellite navigation systems are required to be accurate to provide real-time information. Reliable augmentation of global navigation satellite systems has been made possible through satellite-based augmentation systems, ground-based augmentation and additional navigational sensors like automated celestial navigation systems, inertial navigation systems etc. This chapter provides comprehensive information on the classification of satellite navigation systems.

## GNSS Augmentation

Augmentation of a global navigation satellite system (GNSS) is a method of improving the navigation system's attributes, such as accuracy, reliability, and availability, through the integration of external information into the calculation process. There are many such systems in place and they are generally named or described based on how the GNSS sensor receives the external information. Some systems transmit additional information about sources of error (such as clock drift, ephemeris, or ionospheric delay), others provide direct measurements of how much the signal was off in the past, while a third group provide additional vehicle information to be integrated in the calculation process.

### Satellite-based Augmentation System

A satellite-based augmentation system (SBAS) is a system that supports wide-area or regional augmentation through the use of additional satellite-broadcast messages. Such systems are commonly composed of multiple ground stations, located at accurately-surveyed points. The ground stations take measurements of one or more of the GNSS satellites, the satellite signals, or other environmental factors which may impact the signal received by the users. Using these measurements, information messages are created and sent to one or more satellites for broadcast to the end users. SBAS is sometimes synonymous with WADGPS, wide-area DGPS.

### Implementations

For details on how various SBAS are implemented, please see the following articles:

- The Wide Area Augmentation System (WAAS), operated by the United States

Federal Aviation Administration (FAA).

- The European Geostationary Navigation Overlay Service (EGNOS), operated by the ESSP (on behalf of EU's GSA).

- The Multi-functional Satellite Augmentation System (MSAS) system, operated by Japan's Ministry of Land, Infrastructure and Transport Japan Civil Aviation Bureau (JCAB).

- The Quasi-Zenith Satellite System (QZSS), proposed by Japan.

- The GPS Aided Geo Augmented Navigation (GAGAN) system being operated by India.

- The GLONASS System for Differential Correction and Monitoring (SDCM), proposed by Russia.

- The Satellite Navigation Augmentation System (SNAS), proposed by China.

- The Wide Area GPS Enhancement (WAGE), operated by the United States Department of Defense for use by military and authorized receivers.

- The commercial StarFire navigation system, operated by John Deere.

- The commercial Starfix DGPS System and OmniSTAR system, operated by Fugro.

- The GPS·C, short for GPS Correction, was a Differential GPS data source for most of Canada maintained by the Canadian Active Control System, part of Natural Resources Canada - now decommissioned.

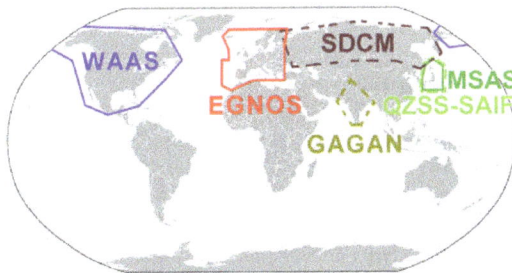

Service areas of satellite-based augmentation systems (SBAS).

## Ground-based Augmentation System

Each of the terms ground-based augmentation system (GBAS) and ground-based regional augmentation system (GRAS) describe a system that supports augmentation through the use of terrestrial radio messages. As with the *satellite based augmentation systems* detailed above, ground based augmentation systems are commonly composed of one or more accurately surveyed ground stations, which take measurements concerning the GNSS, and one or more radio transmitters, which transmit the information directly to the end user from the ground up thus avoiding the constraints associated with GEO Satellites at high latitudes.

Generally, GBAS is localized, supporting receivers within 23 nautical miles, and transmitting in the very high frequency (VHF) band.

The shorter the distance between the ground station that calculates the differential corrections to the inbound plane, the higher the accuracy is likely to be. There are stricter Safety requirements on GBAS systems relative to SBAS systems since GBAS is intended mainly for the landing phase where real-time accuracy and signal integrity control is critical, especially when weather deteriorates to the extent that there is no visibility (CAT-I/II/III conditions) for which SBAS is not intended or suitable.

### Various Ground-based Augmentation Systems

- International Civil Aviation Organization Ground-Based Augmentation System (GBAS) applies to precision approach landing of civil aircraft. Originally this system was called the Local Area Augmentation System (LAAS)

- The US Nationwide Differential GPS System (NDGPS), An augmentation system for users on U.S. land and waterways.

### Additional Navigation Sensors

The augmentation may also take the form of additional information being blended into the position calculation. Many times the additional avionics operate via separate principles than the GNSS and are not necessarily subject to the same sources of error or interference. A system such as this is referred to as an aircraft-based augmentation system (ABAS) by the ICAO.

The additional sensors may include:

- eLORAN receivers

- Automated Celestial navigation systems

- Inertial Navigation Systems

- Simple Dead reckoning systems (composed of a gyro compass and a distance measurement)

## Wide Area Augmentation System

The Wide Area Augmentation System (WAAS) is an air navigation aid developed by the Federal Aviation Administration (prime contractor Raytheon Company) to augment the Global Positioning System (GPS), with the goal of improving its accuracy, integrity, and availability. Essentially, WAAS is intended to enable aircraft to rely on GPS for all phases of flight, including precision approaches to any airport within its coverage area.

FAA WAAS logo

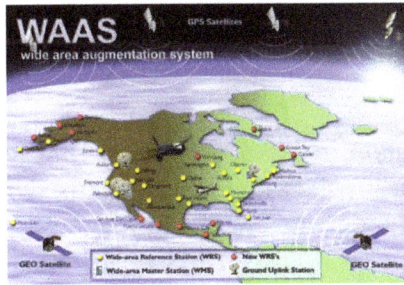

WAAS system overview

WAAS uses a network of ground-based reference stations, in North America and Hawaii, to measure small variations in the GPS satellites' signals in the western hemisphere. Measurements from the reference stations are routed to master stations, which queue the received Deviation Correction (DC) and send the correction messages to geostationary WAAS satellites in a timely manner (every 5 seconds or better). Those satellites broadcast the correction messages back to Earth, where WAAS-enabled GPS receivers use the corrections while computing their positions to improve accuracy.

The International Civil Aviation Organization (ICAO) calls this type of system a satellite-based augmentation system (SBAS). Europe and Asia are developing their own SBASs, the Indian GPS Aided Geo Augmented Navigation (GAGAN), the European Geostationary Navigation Overlay Service (EGNOS) and the Japanese Multi-functional Satellite Augmentation System (MSAS), respectively. Commercial systems include StarFire and OmniSTAR.

## WAAS Objectives

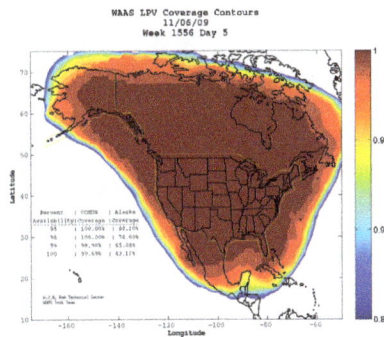

Typical WAAS service area. Dark red indicates best WAAS coverage. The service contours change over time with satellite geometry and ionospheric conditions.

## Accuracy

The WAAS specification requires it to provide a position accuracy of 7.6 metres (25 ft) or better (for both lateral and vertical measurements), at least 95% of the time. Actual performance measurements of the system at specific locations have shown it typically provides better than 1.0 metre (3 ft 3 in) laterally and 1.5 metres (4 ft 11 in) vertically throughout most of the contiguous United States and large parts of Canada and Alaska. With these results, WAAS is capable of achieving the required Category I precision approach accuracy of 16 metres (52 ft) laterally and 4.0 metres (13.1 ft) vertically.

## Integrity

Integrity of a navigation system includes the ability to provide timely warnings when its signal is providing misleading data that could potentially create hazards. The WAAS specification requires the system detect errors in the GPS or WAAS network and notify users within 6.2 seconds. Certifying that WAAS is safe for instrument flight rules (IFR) (i.e. flying in the clouds) requires proving there is only an extremely small probability that an error exceeding the requirements for accuracy will go undetected. Specifically, the probability is stated as $1 \times 10^{-7}$, and is equivalent to no more than 3 seconds of bad data per year. This provides integrity information equivalent to or better than Receiver Autonomous Integrity Monitoring (RAIM).

## Availability

Availability is the probability that a navigation system meets the accuracy and integrity requirements. Before the advent of WAAS, GPS specifications allowed for system unavailability for as much as a total time of four days per year (99% availability). The WAAS specification mandates availability as 99.999% (five nines) throughout the service area, equivalent to a downtime of just over 5 minutes per year.

## Operation

WAAS reference station in Barrow, Alaska

As with GPS in general, WAAS is composed of three main segments: the ground segment, space segment, and user segment.

## Ground Segment

The ground segment is composed of multiple Wide-area Reference Stations (WRS). These precisely surveyed ground stations monitor and collect information on the GPS signals, then send their data to three Wide-area Master Stations (WMS) using a terrestrial communications network. The reference stations also monitor signals from WAAS geostationary satellites, providing integrity information regarding them as well. As of October 2007 there were 38 WRSs: twenty in the contiguous United States (CONUS), seven in Alaska, one in Hawaii, one in Puerto Rico, five in Mexico, and four in Canada.

Using the data from the WRS sites, the WMSs generate two different sets of corrections: fast and slow. The fast corrections are for errors which are changing rapidly and primarily concern the GPS satellites' instantaneous positions and clock errors. These corrections are considered user position-independent, which means they can be applied instantly by any receiver inside the WAAS broadcast footprint. The slow corrections include long-term ephemeric and clock error estimates, as well as ionospheric delay information. WAAS supplies delay corrections for a number of points (organized in a grid pattern) across the WAAS service area.

Once these correction messages are generated, the WMSs send them to two pairs of Ground Uplink Stations (GUS), which then transmit to satellites in the Space segment for rebroadcast to the User segment.

## Reference Stations

Each FAA Air Route Traffic Control Center in the 50 states has a WAAS reference station, except for Indianapolis. There are also stations positioned in Canada, Mexico and Puerto Rico.

## Space Segment

The space segment consists of multiple communication satellites which broadcast the correction messages generated by the WAAS Master Stations for reception by the user segment. The satellites also broadcast the same type of range information as normal GPS satellites, effectively increasing the number of satellites available for a position fix. The space segment consists of three commercial satellites: *Inmarsat-4 F3*, Telesat's *Anik F1R*, and Intelsat's *Galaxy 15*.

The original two WAAS satellites, named *Pacific Ocean Region* (POR) and *Atlantic Ocean Region-West* (AOR-W), were leased space on Inmarsat III satellites. These satellites ceased WAAS transmissions on July 31, 2007. With the end of the Inmarsat lease approaching, two new satellites (Galaxy 15 and Anik F1R) were launched in late 2005. Galaxy 15 is a PanAmSat, and Anik F1R is a Telesat. As with the previous satellites, these are leased services under the FAA's Geostationary Satellite Communications Control Segment contract with Lockheed Martin for WAAS geostationary satellite leased services, who is contracted to provide up to three satellites through the year 2016. A third satellite was later added to the system. From March to November 2010, the FAA broadcast a WAAS test signal on a leased transponder on the Inmarsat-4 F3 satellite. The test signal was not usable for navigation, but could be received and was reported with the identification numbers PRN 133 (NMEA #46). In November 2010, the signal was certified as operational and made available for navigation.

| Satellite Name & Details | PRN | NMEA | Location |
|---|---|---|---|
| Inmarsat-4 F3 | 133 | 46 | 98°W |
| Galaxy 15 | 135 | 48 | 133°W |
| Anik F1R | 138 | 51 | 107.3°W |
| Pacific Ocean Region (POR) *Ceased WAAS transmissions* | 134 | 47 | 178°E |
| Atlantic Ocean Region-West *Ceased WAAS transmissions* | 122 | 35 | 54°W, later moved to 142°W |

In the table above, PRN is the satellite's actual Pseudo-Random Noise code. NMEA is the satellite number sent by some receivers when outputting satellite information. ( NMEA = PRN - 87 ).

EUTELSAT 117 West B, launched on 15 June 2016, also has a WAAS transmitter.

## User Segment

The user segment is the GPS and WAAS receiver, which uses the information broadcast from each GPS satellite to determine its location and the current time, and receives the WAAS corrections from the Space segment. The two types of correction messages received (fast and slow) are used in different ways.

The GPS receiver can immediately apply the fast type of correction data, which includes the corrected satellite position and clock data, and determines its current location using normal GPS calculations. Once an approximate position fix is obtained the receiver begins to use the slow corrections to improve its accuracy. Among the slow correction data is the ionospheric delay. As the GPS signal travels from the satellite to the receiver, it passes through the ionosphere. The receiver calculates the location where the signal pierced the ionosphere and, if it has received an ionospheric delay value for that location, corrects for the error the ionosphere created.

While the slow data can be updated every minute if necessary, ephemeris errors and ionosphere errors do not change this frequently, so they are only updated every two minutes and are considered valid for up to six minutes.

## History and Development

The WAAS was jointly developed by the United States Department of Transportation (DOT) and the Federal Aviation Administration (FAA) as part of the Federal Radio-navigation Program (DOT-VNTSC-RSPA-95-1/DOD-4650.5), beginning in 1994, to provide performance comparable to category 1 instrument landing system (ILS) for all aircraft possessing the appropriately certified equipment. Without WAAS, ionospheric disturbances, clock drift, and satellite orbit errors create too much error and uncertainty in the GPS signal to meet the requirements for a precision approach. A precision approach includes altitude information and provides course guidance, distance from the runway, and elevation information at all points along the approach, usually down to lower altitudes and weather minimums than non-precision approaches.

Prior to the WAAS, the U.S. National Airspace System (NAS) did not have the ability to provide lateral and vertical navigation for precision approaches for all users at all locations. The traditional system for precision approaches is the instrument landing system (ILS), which used a series of radio transmitters each broadcasting a single signal to the aircraft. This complex series of radios needs to be installed at every runway end, some offsite, along a line extended from the runway centerline, making the implementation of a precision approach both difficult and very expensive. The ILS system is composed of 180 different transmitting antennas at each point built. The newer system is free of huge antenna systems at each airport.

For some time the FAA and NASA developed a much improved system, the microwave landing system (MLS). The entire MLS system for a particular approach was isolated in one or two boxes located beside the runway, dramatically reducing the cost of implementation. MLS also offered a number of practical advantages that eased traffic considerations, both for aircraft and radio channels. Unfortunately, MLS would also require every airport and aircraft to upgrade their equipment.

During the development of MLS, consumer GPS receivers of various quality started appearing. GPS offered a huge number of advantages to the pilot, combining all of an aircraft's long-distance navigation systems into a single easy-to-use system, often small enough to be hand held. Deploying an aircraft navigation system based on GPS was largely a problem of developing new techniques and standards, as opposed to new equipment. The FAA started planning to shut down their existing long-distance systems (VOR and NDBs) in favor of GPS. This left the problem of approaches, however. GPS is simply not accurate enough to replace ILS systems. Typical accuracy is about 15 metres (49 ft), whereas even a "CAT I" approach, the least demanding, requires a vertical accuracy of 4 metres (13 ft).

This inaccuracy in GPS is mostly due to large "billows" in the ionosphere, which slow the radio signal from the satellites by a random amount. Since GPS relies on timing the signals to measure distances, this slowing of the signal makes the satellite appear farther away. The billows move slowly, and can be characterized using a variety of methods from the ground, or by examining the GPS signals themselves. By broadcasting this information to GPS receivers every minute or so, this source of error can be significantly reduced. This led to the concept of Differential GPS, which used separate radio systems to broadcast the correction signal to receivers. Aircraft could then install a receiver which would be plugged into the GPS unit, the signal being broadcast on a variety of frequencies for different users (FM radio for cars, longwave for ships, etc.). Broadcasters of the required power generally cluster around larger cities, making such DGPS systems less useful for wide-area navigation. Additionally, most radio signals are either line-of-sight, or can be distorted by the ground, which made DGPS difficult to use as a precision approach system or when flying low for other reasons.

The FAA considered systems that could allow the same correction signals to be broadcast over a much wider area, such as from a satellite, leading directly to WAAS. Since a GPS unit already consists of a satellite receiver, it made much more sense to send out the correction signals on the same frequencies used by GPS units, than to use an entirely separate system and thereby double the probability of failure. In addition to lowering implementation costs by "piggybacking" on a planned satellite launch, this also allowed the signal to be broadcast from geostationary orbit, which meant a small number of satellites could cover all of North America.

On July 10, 2003, the WAAS signal was activated for general aviation, covering 95% of the United States, and portions of Alaska offering 350 feet (110 m) minimums.

On January 17, 2008, Alabama-based Hickok & Associates became the first designer of helicopter WAAS with Localizer Performance (LP) and Localizer Performance with Vertical guidance (LPV) approaches, and the only entity with FAA-approved criteria (which even FAA has yet to develop). This helicopter WAAS criteria offers as low as 250 foot minimums and decreased visibility requirements to enable missions previously not possible. On April 1, 2009, FAA AFS-400 approved the first three helicopter WAAS GPS approach procedures for Hickok & Associates' customer California Shock/Trauma Air Rescue (CALSTAR). Since then they have designed many approved WAAS helicopter approaches for various EMS hospitals and air providers, within the United States as well as in other countries and continents.

On December 30, 2009, Seattle-based Horizon Air flew the first scheduled-passenger service flight using WAAS with LPV on flight 2014, a Portland to Seattle flight operated by a Bombardier Q400 with a WAAS FMS from Universal Avionics. The airline, in partnership with the FAA, will outfit seven Q400-aircraft with WAAS and share flight data to better determine the suitability of WAAS in scheduled air service applications.

# Timeline

## Wide-area Augmentation System (WAAS) Timeline

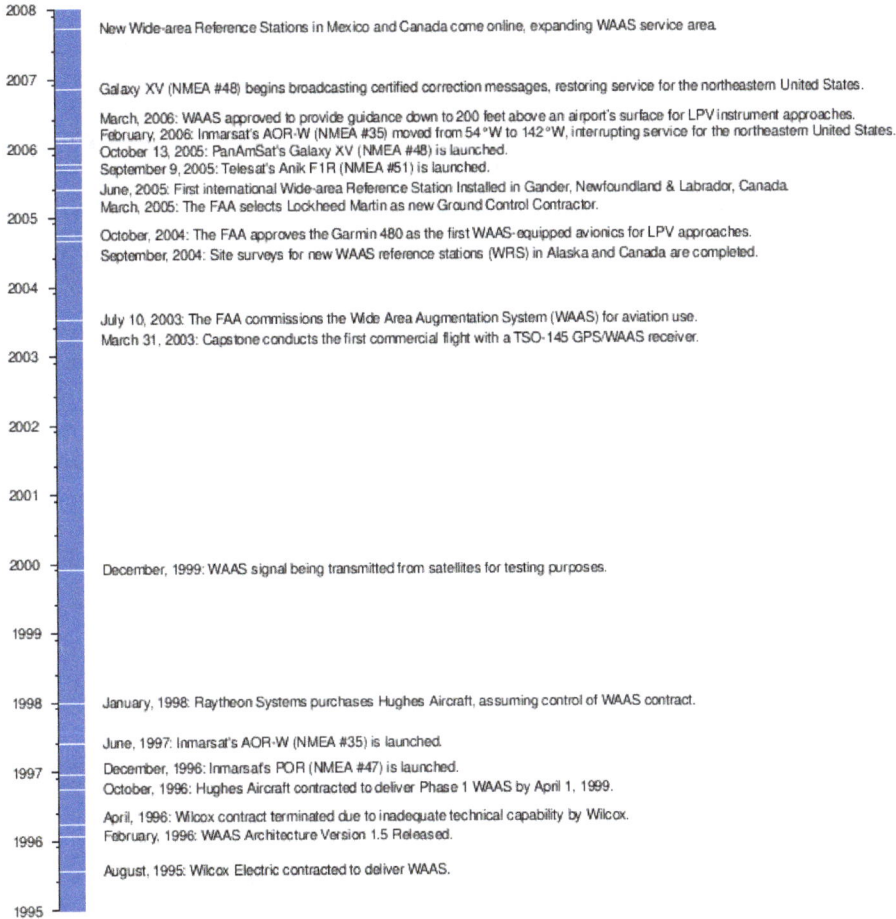

|  |  |
|------|-------|
| 2008 | New Wide-area Reference Stations in Mexico and Canada come online, expanding WAAS service area. |
| 2007 | Galaxy XV (NMEA #48) begins broadcasting certified correction messages, restoring service for the northeastern United States. |
| 2006 | March, 2006: WAAS approved to provide guidance down to 200 feet above an airport's surface for LPV instrument approaches. |
| | February, 2006: Inmarsat's AOR-W (NMEA #35) moved from 54°W to 142°W, interrupting service for the northeastern United States. |
| | October 13, 2005: PanAmSat's Galaxy XV (NMEA #48) is launched. |
| | September 9, 2005: Telesat's Anik F1R (NMEA #51) is launched. |
| 2005 | June, 2005: First international Wide-area Reference Station installed in Gander, Newfoundland & Labrador, Canada. |
| | March, 2005: The FAA selects Lockheed Martin as new Ground Control Contractor. |
| | October, 2004: The FAA approves the Garmin 480 as the first WAAS-equipped avionics for LPV approaches. |
| | September, 2004: Site surveys for new WAAS reference stations (WRS) in Alaska and Canada are completed. |
| 2003 | July 10, 2003: The FAA commissions the Wide Area Augmentation System (WAAS) for aviation use. |
| | March 31, 2003: Capstone conducts the first commercial flight with a TSO-145 GPS/WAAS receiver. |
| 2000 | December, 1999: WAAS signal being transmitted from satellites for testing purposes. |
| 1998 | January, 1998: Raytheon Systems purchases Hughes Aircraft, assuming control of WAAS contract. |
| 1997 | June, 1997: Inmarsat's AOR-W (NMEA #35) is launched. |
| | December, 1996: Inmarsat's POR (NMEA #47) is launched. |
| | October, 1996: Hughes Aircraft contracted to deliver Phase 1 WAAS by April 1, 1999. |
| 1996 | April, 1996: Wilcox contract terminated due to inadequate technical capability by Wilcox. |
| | February, 1996: WAAS Architecture Version 1.5 Released. |
| 1995 | August, 1995: Wilcox Electric contracted to deliver WAAS. |

# Comparison of Accuracy

| A comparison of various radionavigation system accuracies | | |
|---|---|---|
| **System** | **95% Accuracy (Lateral / Vertical)** | **Details** |
| LORAN-C Specification | 460 m / 460 m | The specified absolute accuracy of the LORAN-C system. |
| Distance Measuring Equipment (DME) Specification | 185 m (Linear) | DME is a radionavigation aid that can calculate the linear distance from an aircraft to ground equipment. |
| GPS Specification | 100 m / 150 m | The specified accuracy of the GPS system with the Selective Availability (SA) option turned on. SA was employed by the U.S. Government until May 1, 2000. |

| | | |
|---|---|---|
| LORAN-C Measured Repeatability | 50 m / 50 m | The U.S. Coast Guard reports "return to position" accuracies of 50 meters in time difference mode. |
| eLORAN Repeatability | | Modern LORAN-C receivers, which use all the available signals simultaneously and H-field antennas. |
| Differential GPS (DGPS) | 10 m / 10 m | This is the Differential GPS (DGPS) worst-case accuracy. According to the 2001 Federal Radionavigation Systems (FRS) report published jointly by the U.S. DOT and Department of Defense (DoD), accuracy degrades with distance from the facility; it can be < 1 m but will normally be < 10 m. |
| Wide Area Augmentation System (WAAS) Specification | 7.6 m / 7.6 m | The worst-case accuracy that the WAAS must provide to be used in precision approaches. |
| GPS Measured | 2.5 m / 4.7 m | The actual measured accuracy of the system (excluding receiver errors), with SA turned off, based on the findings of the FAA's National Satellite Test Bed, or NSTB. |
| WAAS Measured | 0.9 m / 1.3 m | The actual measured accuracy of the system (excluding receiver errors), based on the NSTB's findings. |
| Local Area Augmentation System (LAAS) Specification | | The goal of the LAAS program is to provide Category IIIC ILS capability. This will allow aircraft to land with zero visibility utilizing 'autoland' systems and will indicate a very high accuracy of < 1 m. |

## Benefits

*GUS Facility in Napa, California*
WAAS ground uplink station (GUS) in Napa, California

WAAS addresses all of the "navigation problem", providing highly accurate positioning that is extremely easy to use, for the cost of a single receiver installed on the aircraft. Ground- and space-based infrastructure is relatively limited, and no on-airport system is needed. WAAS allows a precision approach to be published for any airport, for the cost of developing the procedures and publishing the new approach plates. This means that almost any airport can have a precision approach and the cost of implementation is dramatically reduced.

Additionally WAAS works just as well between airports. This allows the aircraft to fly directly from one airport to another, as opposed to following routes based on ground-based signals. This can cut route distances considerably in some cases, saving both time and fuel. In addition, because of its ability to provide information on the accuracy of each GPS satellite's information, aircraft equipped with WAAS are permitted to fly at lower en-route altitudes than was possible with ground-based systems, which were often blocked by terrain of varying elevation. This enables pilots to safely fly at lower altitudes, not having to rely on ground-based systems. For unpressurized aircraft, this conserves oxygen and enhances safety.

The above benefits create not only convenience, but also have the potential to generate significant cost savings. The cost to provide the WAAS signal, serving all 5,400 public use airports, is just under US$50 million per year. In comparison, the current ground based systems such as the Instrument Landing System (ILS), installed at only 600 airports, cost US$82 million in annual maintenance. Without ground navigation hardware to purchase, the total cost of publishing a runway's WAAS approach is approximately US$50,000; compared to the $1,000,000 to $1,500,000 cost to install an ILS radio system.

Further savings can come from the nighttime closure of airport towers with a low volume of traffic. The FAA is reviewing 48 towers for such a potential reduction of services, which it estimates will save around US$100,000 per year at each tower, for a total annual savings of nearly US$5 million.

## Drawbacks and Limitations

For all its benefits, WAAS is not without drawbacks and critical limitations:

- Space weather. All man-made satellite systems are subject to space weather and space debris threats. For example, a solar super-storm event composed of an extremely large and fast earthbound Coronal Mass Ejection (CME) could disable the geosynchronous or GPS satellite elements of WAAS.

- The broadcasting satellites are geostationary, which causes them to be less than 10° above the horizon for locations north of 71.4° latitude. This means aircraft in areas of Alaska or northern Canada may have difficulty maintaining a lock on the WAAS signal.

- To calculate an ionospheric grid point's delay, that point must be located between a satellite and a reference station. The low number of satellites and ground stations limit the number of points which can be calculated.

- Aircraft conducting WAAS approaches must possess certified GPS receivers, which are much more expensive than non-certified units. In 2006, Garmin's least expensive certified receiver, the GNS 430W, had a suggested retail price of US$10,750.

- WAAS is not capable of the accuracies required for Category II or III ILS approaches. Thus, WAAS is not a sole-solution and either existing ILS equipment must be maintained or it must be replaced by new systems, such as the Local Area Augmentation System (LAAS).

- WAAS Localizer Performance with Vertical guidance (LPV) approaches with 200-foot minimums will not be published for airports without medium intensity lighting, precision runway markings and a parallel taxiway. Smaller airports, which currently may not have these features, would have to upgrade their facilities or require pilots to use higher minimums.

- As precision increases and error approaches zero, the navigation paradox states that there is an increased collision risk, as the likelihood of two craft occupying the same space on the shortest distance line between two navigational points has increased.

## Future of WAAS

## Improvement to Aviation Operations

In 2007, WAAS vertical guidance was projected to be available nearly all the time (greater than 99%), and its coverage encompasses the full continental U.S., most of Alaska, northern Mexico, and southern Canada. At that time, the accuracy of WAAS would meet or exceed the requirements for Category 1 ILS approaches, namely, three-dimensional position information down to 200 feet (60 m) above touchdown zone elevation.

## Software Improvements

Software improvements, to be implemented by September 2008, significantly improve signal availability of vertical guidance throughout the CONUS and Alaska. Area covered by the 95% available LPV solution in Alaska improves from 62% to 86%. And in the CONUS, the 100% availability LPV-200 coverage rises from 48% to 84%, with 100% coverage of the LPV solution.

## Space Segment Upgrades

Both Galaxy XV (PRN #135) and Anik F1R (PRN #138) contain an L1 & L5 GPS payload. This means they will potentially be usable with the L5 modernized GPS signals when the new signals and receivers become available. With L5, avionics will be able to use a combination of signals to provide the most accurate service possible, thereby increasing availability of the service. These avionics systems will use ionospheric corrections broadcast by WAAS, or self-generated onboard dual frequency corrections, depending on which one is more accurate.

# Multi-functional Transport Satellite

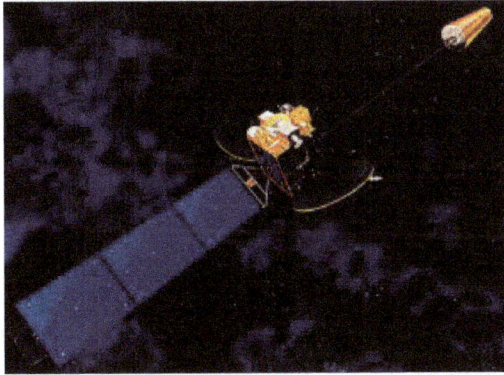

MTSAT-1 *Himawari 6*

Multifunctional Transport Satellites (MTSAT) are a series of weather and aviation control satellites. They are geostationary satellites owned and operated by the Japanese Ministry of Land, Infrastructure, Transport and Tourism and the Japan Meteorological Agency (JMA), and provide coverage for the hemisphere centred on 140° East; this includes Japan and Australia who are the principal users of the satellite imagery that MTSAT provides. They replace the GMS-5 satellite, also known as *Himawari 5* ("himawari" or "ひまわり" meaning "sunflower"). They can provide imagery in five wavelength bands — visible and four infrared, including the water vapour channel. The visible light camera has a resolution of 1 km; the infrared cameras have 4 km (resolution is lower away from the equator at 140° East). The spacecraft have a planned lifespan of five years. MTSAT-1 and 1R were built by Space Systems/Loral. MTSAT-2 was built by Mitsubishi. It was replaced by Himawari 8 on 7 July 2015.

## MTSAT-1 and GOES-9

## Launch Failure

The launch of MTSAT-1, on a Japanese H-II rocket, failed on November 15, 1999 and the spacecraft was destroyed. GMS-5, the satellite MTSAT-1 was intended to replace, was decommissioned on April 1, 2003 leaving Japan without weather satellite imagery.

## NOAA Loan

To fill in the void, The United States National Oceanic and Atmospheric Administration (NOAA) loaned the GOES-9 satellite to the JMA and repositioned it over 155° East on May 22, 2003.

## MTSAT-1R

MTSAT-1R (also known as Himawari 6) was successfully launched on a H-IIA on

February 26, 2005 and became partially operational on June 28, 2005 — the aviation payload was not functional as two MTSATs are required for air traffic control. GOES-9 was decommissioned when MTSAT-1R came online in June 2005. Its solar sail counteracts the torque produced by sunlight pressure on the solar array. The trim tab on the solar array makes small adjustments to the torque balance.

MTSAT-1R was decommissioned on December 4, 2015, due to fuel limitations.

## MTSAT-2

MTSAT-2 (also known as Himawari 7) successfully launched on February 18, 2006 and is positioned at 145° East. The weather functions of MTSAT-2 were put into hibernation until the end of MTSAT-1R's life (5 years from launch). The transportation and communication functions of MTSAT-2 will be utilized prior to that time.

## Attitude Control Malfunction

On November 5, 2007 JMA announced a malfunction in the attitude control of MTSAT-2. Attitude control was restored November 7, 2007. The presumed cause of the malfunction was improper functioning of an attitude control thruster. A spare thruster was used to return the spacecraft to normal operation.

## Differential GPS

Transportable DGPS reference station *Baseline HD* by CLAAS for use in satellite-assisted steering systems in modern agriculture

Differential Global Positioning System (DGPS) is an enhancement to Global Positioning System that provides improved location accuracy, from the 15-meter nominal GPS accuracy to about 10 cm in case of the best implementations.

DGPS uses a network of fixed, ground-based reference stations to broadcast the difference between the positions indicated by the GPS satellite systems and the known fixed positions. These stations broadcast the difference between the measured satellite pseudoranges and actual (internally computed) pseudoranges, and receiver stations may correct their pseudoranges by the same amount. The digital correction signal is typically broadcast locally over ground-based transmitters of shorter range.

The term refers to a general technique of augmentation. The United States Coast Guard (USCG) and Canadian Coast Guard (CCG) each run such systems in the U.S. and Canada on the longwave radio frequencies between 285 kHz and 325 kHz near major waterways and harbors. The USCG's DGPS system has been named NDGPS (Nationwide DGPS) and is now jointly administered by the Coast Guard and the U.S. Department of Transportation's Federal Highway Administration. It consists of broadcast sites located throughout the inland and coastal portions of the United States including Alaska, Hawaii and Puerto Rico.

A similar system that transmits corrections from orbiting satellites instead of ground-based transmitters is called a Wide-Area DGPS (WADGPS) or Satellite Based Augmentation System.

## History

When GPS was first being put into service, the US military was concerned about the possibility of enemy forces using the globally available GPS signals to guide their own weapon systems. Originally, the government thought the "coarse acquisition" (C/A) signal would only give about 100 meter accuracy, but with improved receiver designs, the actual accuracy was 20 to 30 meters. Starting in March 1990, to avoid providing such unexpected accuracy, the C/A signal transmitted on the L1 frequency (1575.42 MHz) was deliberately degraded by offsetting its clock signal by a random amount, equivalent to about 100 meters of distance. This technique, known as "Selective Availability", or SA for short, seriously degraded the usefulness of the GPS signal for non-military users. More accurate guidance was possible for users of dual frequency GPS receivers that also received the L2 frequency (1227.6 MHz), but the L2 transmission, intended for military use, was encrypted and was only available to authorised users with the encryption keys.

This presented a problem for civilian users who relied upon ground-based radio navigation systems such as LORAN, VOR and NDB systems costing millions of dollars each year to maintain. The advent of a global navigation satellite system (GNSS) could provide greatly improved accuracy and performance at a fraction of the cost. The accuracy inherent in the S/A signal was however too poor to make this realistic. The military received multiple requests from the Federal Aviation Administration (FAA), United States Coast Guard (USCG) and United States Department of Transportation (DOT) to set S/A aside to enable civilian use of GNSS, but remained steadfast in its objection on grounds of security.

Through the early to mid 1980s, a number of agencies developed a solution to the SA "problem". Since the SA signal was changed slowly, the effect of its offset on positioning was relatively fixed – that is, if the offset was "100 meters to the east", that offset would be true over a relatively wide area. This suggested that broadcasting this offset to local GPS receivers could eliminate the effects of SA, resulting in measurements closer to GPS's theoretical performance, around 15 meters. Additionally, another major source of errors in a GPS fix is due to transmission delays in the ionosphere, which could also be measured and corrected for in the broadcast. This offered an improvement to about 5 meters accuracy, more than enough for most civilian needs.

The US Coast Guard was one of the more aggressive proponents of the DGPS system, experimenting with the system on an ever-wider basis through the late 1980s and early 1990s. These signals are broadcast on marine longwave frequencies, which could be received on existing radiotelephones and fed into suitably equipped GPS receivers. Almost all major GPS vendors offered units with DGPS inputs, not only for the USCG signals, but also aviation units on either VHF or commercial AM radio bands.

They started sending out "production quality" DGPS signals on a limited basis in 1996, and rapidly expanded the network to cover most US ports of call, as well as the Saint Lawrence Seaway in partnership with the Canadian Coast Guard. Plans were put into place to expand the system across the US, but this would not be easy. The quality of the DGPS corrections generally fell with distance, and large transmitters capable of covering large areas tend to cluster near cities. This meant that lower-population areas, notably in the midwest and Alaska, would have little coverage by ground-based GPS. As of November 2013 the USCG's national DGPS system comprises 85 broadcast sites which provide dual coverage to almost the entire US coastline and inland navigable waterways including Alaska, Hawaii, and Puerto Rico. In addition the system provides single or dual coverage to a majority of the inland portion of United States. Instead, the FAA (and others) started studying broadcasting the signals across the entire hemisphere from communications satellites in geostationary orbit. This led to the Wide Area Augmentation System (WAAS) and similar systems, although these are generally not referred to as DGPS, or alternatively, "wide-area DGPS". WAAS offers accuracy similar to the USCG's ground-based DGPS networks, and there has been some argument that the latter will be turned off as WAAS becomes fully operational.

By the mid-1990s it was clear that the SA system was no longer useful in its intended role. DGPS would render it ineffective over the US, precisely where it was considered most needed. Additionally, experience during the Gulf War demonstrated that the widespread use of civilian receivers by U.S. forces meant that leaving SA turned on was thought to harm the U.S. more than if it were turned off. After many years of pressure, it took an executive order by President Bill Clinton to get SA turned off permanently in 2000.

Nevertheless, by this point DGPS had evolved into a system for providing more accuracy than even a non-SA GPS signal could provide on its own. There are several other sources of error that share the same characteristics as SA in that they are the same over large areas and for "reasonable" amounts of time. These include the ionospheric effects mentioned earlier, as well as errors in the satellite position ephemeris data and clock drift on the satellites. Depending on the amount of data being sent in the DGPS correction signal, correcting for these effects can reduce the error significantly, the best implementations offering accuracies of under 10 cm.

In addition to continued deployments of the USCG and FAA sponsored systems, a number of vendors have created commercial DGPS services, selling their signal (or receivers for it) to users who require better accuracy than the nominal 15 meters GPS offers. Almost all commercial GPS units, even hand-held units, now offer DGPS data inputs, and many also support WAAS directly. To some degree, a form of DGPS is now a natural part of most GPS operations.

## Operation

DGPS Reference Station (choke ring antenna)

A reference station calculates differential corrections for its own location and time. Users may be up to 200 nautical miles (370 km) from the station, however, and some of the compensated errors vary with space: specifically, satellite ephemeris errors and those introduced by ionospheric and tropospheric distortions. For this reason, the accuracy of DGPS decreases with distance from the reference station. The problem can be aggravated if the user and the station lack "inter visibility"—when they are unable to see the same satellites.

## Accuracy

The United States *Federal Radionavigation Plan* and the IALA *Recommendation on the Performance and Monitoring of DGNSS Services in the Band 283.5–325 kHz* cite the United States Department of Transportation's 1993 estimated error growth of 0.67 m per 100 km from the broadcast site but measurements of accuracy across the Atlantic, in Portugal, suggest a degradation of just 0.22 m per 100 km.

## Variations

DGPS can refer to any type of Ground Based Augmentation System (GBAS). There are many operational systems in use throughout the world, according to the US Coast Guard, 47 countries operate systems similar to the US NDGPS (Nationwide Differential Global Positioning System).

A list can be found at World DGPS Database for Dxers

## European DGPS Network

The European DGPS network has been mainly developed by the Finnish and Swedish maritime administrations in order to improve safety in the archipelago between the two countries.

In the UK and Ireland, the system was implemented as a maritime navigation aid to fill the gap left by the demise of the Decca Navigator System in 2000. With a network of 12 transmitters sited around the coastline and three control stations, it was set up in 1998 by the countries' respective General Lighthouse Authorities (GLA) — Trinity House covering England, Wales and the Channel Islands, the Northern Lighthouse Board covering Scotland and the Isle of Man and the Commissioners of Irish Lights, covering the whole of Ireland. Transmitting on the 300 kHz band, the system underwent testing and two additional transmitters were added before the system was declared operational in 2002.

Trinity House - DGNSS Stations: UK and Ireland

Effective Solutions (Data Products) - European Differential Beacon Transmitters - Details and map

## United States NDGPS

The United States Department of Transportation, in conjunction with the Federal Highway Administration, the Federal Railroad Administration and the National Geodetic Survey appointed the Coast Guard as the maintaining agency for the U.S. Nationwide DGPS network (NDGPS). The system is an expansion of the previous Maritime Differential GPS (MDGPS), which the Coast Guard began in the late 1980s and completed in

March 1999. MDGPS only covered coastal waters, the Great Lakes, and the Mississippi River inland waterways, while NDGPS expands this to include complete coverage of the continental United States. The centralized Command and Control unit is the USCG Navigation Center, based in Alexandria, VA. There are currently 85 NDGPS sites in the US network, administered by the U.S. Department of Homeland Security Navigation Center.

## Canadian DGPS

The Canadian system is similar to the US system and is primarily for maritime usage covering the Atlantic and Pacific coast as well as the Great Lakes and Saint Lawrence Seaway.

## Australia

Australia runs three DGPS systems: one is mainly for marine navigation, broadcasting its signal on the longwave band; another is used for land surveys and land navigation, and has corrections broadcast on the Commercial FM radio band. While the third at Sydney airport is currently undergoing testing for precision landing of aircraft (2011), as a backup to the Instrument Landing System at least until 2015. It is called the Ground Based Augmentation System. Corrections to aircraft position are broadcast via the aviation VHF band.

## Post Processing

Post-processing is used in Differential GPS to obtain precise positions of unknown points by relating them to known points such as survey markers.

The GPS measurements are usually stored in computer memory in the GPS receivers, and are subsequently transferred to a computer running the GPS post-processing software. The software computes baselines using simultaneous measurement data from two or more GPS receivers.

The baselines represent a three-dimensional line drawn between the two points occupied by each pair of GPS antennas. The post-processed measurements allow more precise positioning, because most GPS errors affect each receiver nearly equally, and therefore can be cancelled out in the calculations.

Differential GPS measurements can also be computed in real-time by some GPS receivers if they receive a correction signal using a separate radio receiver, for example in Real Time Kinematic (RTK) surveying or navigation.

The improvement of GPS positioning doesn't require simultaneous measurements of two or more receivers in any case, but can also be done by special use of a *single* device. In the 1990s when even handheld receivers were quite expensive, some methods of quasi-differential GPS were developed, using the receiver by quick turns of positions or loops of 3-10 survey points.

## Spacecraft Propulsion

A remote camera captures a close-up view of a Space Shuttle Main Engine during a test firing at the John C. Stennis Space Center in Hancock County, Mississippi.

Spacecraft propulsion is any method used to accelerate spacecraft and artificial satellites. There are many different methods. Each method has drawbacks and advantages, and spacecraft propulsion is an active area of research. However, most spacecraft today are propelled by forcing a gas from the back/rear of the vehicle at very high speed through a supersonic de Laval nozzle. This sort of engine is called a rocket engine.

All current spacecraft use chemical rockets (bipropellant or solid-fuel) for launch, though some (such as the Pegasus rocket and SpaceShipOne) have used air-breathing engines on their first stage. Most satellites have simple reliable chemical thrusters (often monopropellant rockets) or resistojet rockets for orbital station-keeping and some use momentum wheels for attitude control. Soviet bloc satellites have used electric propulsion for decades, and newer Western geo-orbiting spacecraft are starting to use them for north-south stationkeeping and orbit raising. Interplanetary vehicles mostly use chemical rockets as well, although a few have used ion thrusters and Hall effect thrusters (two different types of electric propulsion) to great success.

## Requirements

Artificial satellites must be launched into orbit and once there they must be placed in their nominal orbit. Once in the desired orbit, they often need some form of attitude control so that they are correctly pointed with respect to Earth, the Sun, and possibly some astronomical object of interest. They are also subject to drag from the thin atmosphere, so that to stay in orbit for a long period of time some form of propulsion is occasionally necessary to make small corrections (orbital stationkeeping). Many satellites need to be moved from one orbit to another from time to time, and this also requires propulsion. A satellite's useful life is over once it has exhausted its ability to adjust its orbit.

Spacecraft designed to travel further also need propulsion methods. They need to be launched out of the Earth's atmosphere just as satellites do. Once there, they need to leave orbit and move around.

For interplanetary travel, a spacecraft must use its engines to leave Earth orbit. Once it has done so, it must somehow make its way to its destination. Current interplanetary spacecraft do this with a series of short-term trajectory adjustments. In between these adjustments, the spacecraft simply falls freely along its trajectory. The most fuel-efficient means to move from one circular orbit to another is with a Hohmann transfer orbit: the spacecraft begins in a roughly circular orbit around the Sun. A short period of thrust in the direction of motion accelerates or decelerates the spacecraft into an elliptical orbit around the Sun which is tangential to its previous orbit and also to the orbit of its destination. The spacecraft falls freely along this elliptical orbit until it reaches its destination, where another short period of thrust accelerates or decelerates it to match the orbit of its destination. Special methods such as aerobraking or aerocapture are sometimes used for this final orbital adjustment.

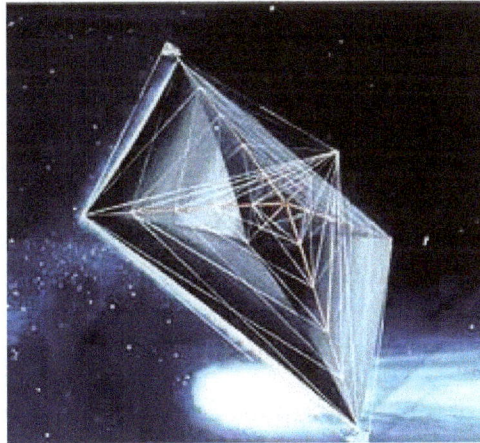
Artist's concept of a solar sail

Some spacecraft propulsion methods such as solar sails provide very low but inexhaustible thrust; an interplanetary vehicle using one of these methods would follow a rather different trajectory, either constantly thrusting against its direction of motion in order to decrease its distance from the Sun or constantly thrusting along its direction of motion to increase its distance from the Sun. The concept has been successfully tested by the Japanese IKAROS solar sail spacecraft.

Spacecraft for interstellar travel also need propulsion methods. No such spacecraft has yet been built, but many designs have been discussed. Because interstellar distances are very great, a tremendous velocity is needed to get a spacecraft to its destination in a reasonable amount of time. Acquiring such a velocity on launch and getting rid of it on arrival will be a formidable challenge for spacecraft designers.

## Effectiveness

When in space, the purpose of a propulsion system is to change the velocity, or $v$, of a spacecraft. Because this is more difficult for more massive spacecraft, designers generally discuss momentum, $mv$. The amount of change in momentum is called impulse. So the goal of a propulsion method in space is to create an impulse.

When launching a spacecraft from Earth, a propulsion method must overcome a higher gravitational pull to provide a positive net acceleration. In orbit, any additional impulse, even very tiny, will result in a change in the orbit path.

The rate of change of velocity is called acceleration, and the rate of change of momentum is called force. To reach a given velocity, one can apply a small acceleration over a long period of time, or one can apply a large acceleration over a short time. Similarly, one can achieve a given impulse with a large force over a short time or a small force over a long time. This means that for maneuvering in space, a propulsion method that produces tiny accelerations but runs for a long time can produce the same impulse as a propulsion method that produces large accelerations for a short time. When launching from a planet, tiny accelerations cannot overcome the planet's gravitational pull and so cannot be used.

Earth's surface is situated fairly deep in a gravity well. The escape velocity required to get out of it is 11.2 kilometers/second. As human beings evolved in a gravitational field of 1g ($9.8 \text{ m/s}^2$), an ideal propulsion system would be one that provides a continuous acceleration of 1g (though human bodies can tolerate much larger accelerations over short periods). The occupants of a rocket or spaceship having such a propulsion system would be free from all the ill effects of free fall, such as nausea, muscular weakness, reduced sense of taste, or leaching of calcium from their bones.

The law of conservation of momentum means that in order for a propulsion method to change the momentum of a space craft it must change the momentum of something else as well. A few designs take advantage of things like magnetic fields or light pressure in order to change the spacecraft's momentum, but in free space the rocket must bring along some mass to accelerate away in order to push itself forward. Such mass is called reaction mass.

In order for a rocket to work, it needs two things: reaction mass and energy. The impulse provided by launching a particle of reaction mass having mass $m$ at velocity $v$ is $mv$. But this particle has kinetic energy $mv^2/2$, which must come from somewhere. In a conventional solid, liquid, or hybrid rocket, the fuel is burned, providing the energy, and the reaction products are allowed to flow out the back, providing the reaction mass.

In an ion thruster, electricity is used to accelerate ions out the back. Here some other source must provide the electrical energy (perhaps a solar panel or a nuclear reactor), whereas the ions provide the reaction mass.

When discussing the efficiency of a propulsion system, designers often focus on effectively using the reaction mass. Reaction mass must be carried along with the rocket and is irretrievably consumed when used. One way of measuring the amount of impulse that can be obtained from a fixed amount of reaction mass is the specific impulse, the impulse per unit weight-on-Earth (typically designated by $I_{sp}$). The unit for this value is seconds. Because the weight on Earth of the reaction mass is often unimportant when discussing vehicles in space, specific impulse can also be discussed in terms of impulse per unit mass. This alternate form of specific impulse uses the same units as velocity (e.g. m/s), and in fact it is equal to the effective exhaust velocity of the engine (typically designated $v_e$). Confusingly, both values are sometimes called specific impulse. The two values differ by a factor of $g_n$, the standard acceleration due to gravity 9.80665 m/s$^2$ ($I_{sp}g_n = v_e$).

A rocket with a high exhaust velocity can achieve the same impulse with less reaction mass. However, the energy required for that impulse is proportional to the exhaust velocity, so that more mass-efficient engines require much more energy, and are typically less energy efficient. This is a problem if the engine is to provide a large amount of thrust. To generate a large amount of impulse per second, it must use a large amount of energy per second. So high-mass-efficient engines require enormous amounts of energy per second to produce high thrusts. As a result, most high-mass-efficient engine designs also provide lower thrust due to the unavailability of high amounts of energy.

## Methods

Propulsion methods can be classified based on their means of accelerating the reaction mass. There are also some special methods for launches, planetary arrivals, and landings.

## Reaction Engines

A reaction engine is an engine which provides propulsion by expelling reaction mass, in accordance with Newton's third law of motion. This law of motion is most commonly paraphrased as: "For every action force there is an equal, but opposite, reaction force".

Examples include both duct engines and rocket engines, and more uncommon variations such as Hall effect thrusters, ion drives and mass drivers. Duct engines are obviously not used for space propulsion due to the lack of air; however some proposed spacecraft have these kinds of engines to assist takeoff and landing.

## Delta-v and Propellant

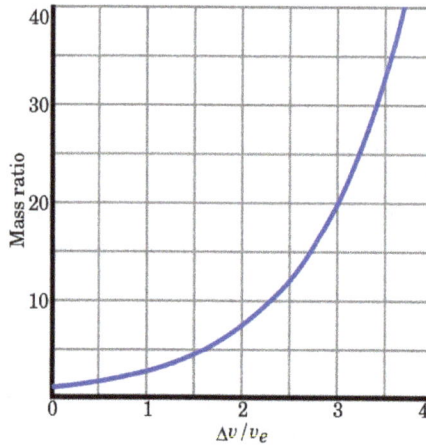

Rocket mass ratios versus final velocity, as calculated from the rocket equation

Exhausting the entire usable propellant of a spacecraft through the engines in a straight line in free space would produce a net velocity change to the vehicle; this number is termed 'delta-v' ($\Delta v$).

If the exhaust velocity is constant then the total $\Delta v$ of a vehicle can be calculated using the rocket equation, where $M$ is the mass of propellant, $P$ is the mass of the payload (including the rocket structure), and $v_e$ is the velocity of the rocket exhaust. This is known as the Tsiolkovsky rocket equation:

$$\Delta v = v_e \ln\left(\frac{M + P}{P}\right).$$

For historical reasons, as discussed above, $v_e$ is sometimes written as

$$v_e = I_{sp} g_o$$

where $I_{sp}$ is the specific impulse of the rocket, measured in seconds, and $g_o$ is the gravitational acceleration at sea level.

For a high delta-v mission, the majority of the spacecraft's mass needs to be reaction mass. Because a rocket must carry all of its reaction mass, most of the initially-expended reaction mass goes towards accelerating reaction mass rather than payload. If the rocket has a payload of mass $P$, the spacecraft needs to change its velocity by $\Delta v$,, and the rocket engine has exhaust velocity $v_e$, then the mass $M$ of reaction mass which is needed can be calculated using the rocket equation and the formula for $I_{sp}$:

$$M = P\left(e^{\frac{\Delta v}{v_e}} - 1\right).$$

For $\Delta v$ much smaller than $v_e$, this equation is roughly linear, and little reaction mass is needed. If $\Delta v$ is comparable to $v_e$, then there needs to be about twice as much fuel as combined payload and structure (which includes engines, fuel tanks, and so on). Beyond this, the growth is exponential; speeds much higher than the exhaust velocity require very high ratios of fuel mass to payload and structural mass.

For a mission, for example, when launching from or landing on a planet, the effects of gravitational attraction and any atmospheric drag must be overcome by using fuel. It is typical to combine the effects of these and other effects into an effective mission delta-v. For example, a launch mission to low Earth orbit requires about 9.3–10 km/s delta-v. These mission delta-vs are typically numerically integrated on a computer.

Some effects such as Oberth effect can only be significantly utilised by high thrust engines such as rockets, i.e. engines that can produce a high g-force (thrust per unit mass, equal to delta-v per unit time).

## Power Use and Propulsive Efficiency

For all reaction engines (such as rockets and ion drives) some energy must go into accelerating the reaction mass. Every engine will waste some energy, but even assuming 100% efficiency, to accelerate an exhaust the engine will need energy amounting to

$$\frac{1}{2}\dot{m}v_e^2$$

This energy is not necessarily lost- some of it usually ends up as kinetic energy of the vehicle, and the rest is wasted in residual motion of the exhaust.

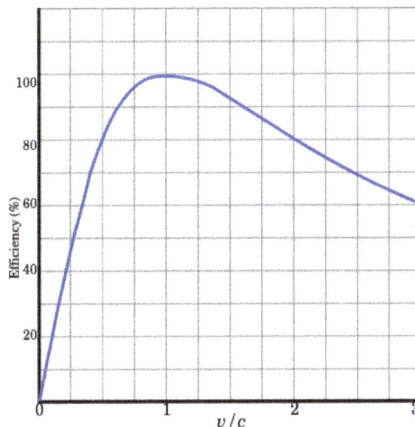

Due to energy carried away in the exhaust, the energy efficiency of a reaction engine varies with the speed of the exhaust relative to the speed of the vehicle, this is called propulsive efficiency

Comparing the rocket equation (which shows how much energy ends up in the final vehicle) and the above equation (which shows the total energy required) shows that even with 100% engine efficiency, certainly not all energy supplied ends up in the vehicle - some of it, indeed usually most of it, ends up as kinetic energy of the exhaust.

The exact amount depends on the design of the vehicle, and the mission. However, there are some useful fixed points:

- if the $I_{sp}$ is fixed, for a mission delta-v, there is a particular $I_{sp}$ that minimises the overall energy used by the rocket. This comes to an exhaust velocity of about ⅔ of the mission delta-v. Drives with a specific impulse that is both high and fixed such as Ion thrusters have exhaust velocities that can be enormously higher than this ideal for many missions.

- if the exhaust velocity can be made to vary so that at each instant it is equal and opposite to the vehicle velocity then the absolute minimum energy usage is achieved. When this is achieved, the exhaust stops in space and has no kinetic energy; and the propulsive efficiency is 100%- all the energy ends up in the vehicle (in principle such a drive would be 100% efficient, in practice there would be thermal losses from within the drive system and residual heat in the exhaust). However, in most cases this uses an impractical quantity of propellant, but is a useful theoretical consideration. Anyway, the vehicle has to move before the method can be applied.

Some drives (such as VASIMR or Electrodeless plasma thruster) actually can significantly vary their exhaust velocity. This can help reduce propellant usage or improve acceleration at different stages of the flight. However the best energetic performance and acceleration is still obtained when the exhaust velocity is close to the vehicle speed. Proposed ion and plasma drives usually have exhaust velocities enormously higher than that ideal (in the case of VASIMR the lowest quoted speed is around 15000 m/s compared to a mission delta-v from high Earth orbit to Mars of about 4000m/s).

It might be thought that adding power generation capacity is helpful, and although initially this can improve performance, this inevitably increases the weight of the power source, and eventually the mass of the power source and the associated engines and propellant dominates the weight of the vehicle, and then adding more power gives no significant improvement.

For, although solar power and nuclear power are virtually unlimited sources of *energy*, the maximum *power* they can supply is substantially proportional to the mass of the powerplant (i.e. specific power takes a largely constant value which is dependent on the particular powerplant technology). For any given specific power, with a large $v_e$ which is desirable to save propellant mass, it turns out that the maximum acceleration is inversely proportional to $v_e$.. Hence the time to reach a required delta-v is proportional to $v_e$. Thus the latter should not be too large.

## Energy

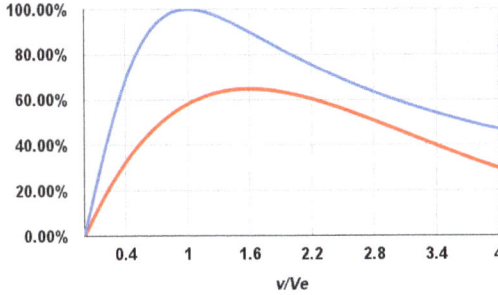

Plot of instantaneous propulsive efficiency (blue) and overall efficiency for a vehicle accelerating from rest (red) as percentages of the engine efficiency

In the ideal case $m_1$ is useful payload and $m_0 - m_1$ is reaction mass (this corresponds to empty tanks having no mass, etc.). The energy required can simply be computed as

$$\frac{1}{2}(m_0 - m_1)v_e^2$$

This corresponds to the kinetic energy the expelled reaction mass would have at a speed equal to the exhaust speed. If the reaction mass had to be accelerated from zero speed to the exhaust speed, all energy produced would go into the reaction mass and nothing would be left for kinetic energy gain by the rocket and payload. However, if the rocket already moves and accelerates (the reaction mass is expelled in the direction opposite to the direction in which the rocket moves) less kinetic energy is added to the reaction mass. To see this, if, for example, =10 km/s and the speed of the rocket is 3 km/s, then the speed of a small amount of expended reaction mass changes from 3 km/s forwards to 7 km/s rearwards. Thus, although the energy required is 50 MJ per kg reaction mass, only 20 MJ is used for the increase in speed of the reaction mass. The remaining 30 MJ is the increase of the kinetic energy of the rocket and payload.

In general:

$$d\left(\frac{1}{2}v^2\right) = vdv = vv_e\frac{dm}{m} = \frac{1}{2}\left[v_e^2 - (v - v_e)^2 + v^2\right]\frac{dm}{m}$$

Thus the specific energy gain of the rocket in any small time interval is the energy gain of the rocket including the remaining fuel, divided by its mass, where the energy gain is equal to the energy produced by the fuel minus the energy gain of the reaction mass. The larger the speed of the rocket, the smaller the energy gain of the reaction mass; if the rocket speed is more than half of the exhaust speed the reaction mass even loses energy on being expelled, to the benefit of the energy gain of the rocket; the larger the speed of the rocket, the larger the energy loss of the reaction mass.

We have

$$\Delta\epsilon = \int vd(\Delta v)$$

where $\dot{o}$ is the specific energy of the rocket (potential plus kinetic energy) and $\Delta v$ is a separate variable, not just the change in $v$. In the case of using the rocket for deceleration, i.e. expelling reaction mass in the direction of the velocity, $v$ should be taken negative.

The formula is for the ideal case again, with no energy lost on heat, etc. The latter causes a reduction of thrust, so it is a disadvantage even when the objective is to lose energy (deceleration).

If the energy is produced by the mass itself, as in a chemical rocket, the fuel value has to be , where for the fuel value also the mass of the oxidizer has to be taken into account. A typical value is v= 4.5 km/s, corresponding to a fuel value of 10.1 MJ/kg. The actual fuel value is higher, but much of the energy is lost as waste heat in the exhaust that the nozzle was unable to extract.

The required energy $E$ is

$$E = \frac{1}{2} m_1 \left( e^{\frac{\Delta v}{v_e}} - 1 \right) v_e^2$$

Conclusions:

- for $\Delta v \ll v_e$ we have $E \approx \frac{1}{2} m_1 v_e \Delta v$

- for a given $\Delta v$,, the minimum energy is needed if $v_e = 0.6275 \Delta v$, requiring an energy of

  $$E = 0.772 m_1 (\Delta v)^2.$$

  In the case of acceleration in a fixed direction, and starting from zero speed, and in the absence of other forces, this is 54.4% more than just the final kinetic energy of the payload. In this optimal case the initial mass is 4.92 times the final mass.

These results apply for a fixed exhaust speed.

Due to the Oberth effect and starting from a nonzero speed, the required potential energy needed from the propellant may be *less* than the increase in energy in the vehicle and payload. This can be the case when the reaction mass has a lower speed after being expelled than before – rockets are able to liberate some or all of the initial kinetic energy of the propellant.

Also, for a given objective such as moving from one orbit to another, the required $\Delta v$ may depend greatly on the rate at which the engine can produce $\Delta v$ and maneuvers may even be impossible if that rate is too low. For example, a launch to Low Earth Orbit (LEO) normally requires a $\Delta v$ of ca. 9.5 km/s (mostly for the speed to be acquired), but if the engine could produce $\Delta v$ at a rate of only slightly more than $g$, it would be a slow launch requiring altogether a very large $\Delta v$ (think of hovering without making any progress in speed or altitude, it would cost a $\Delta v$ of 9.8 m/s each second). If the possible rate is only $g$ or less, the maneuver can not be carried out at all with this engine.

The power is given by

$$P = \frac{1}{2}mav_e = \frac{1}{2}Fv_e$$

where $F$ is the thrust and $a$ the acceleration due to it. Thus the theoretically possible thrust per unit power is 2 divided by the specific impulse in m/s. The thrust efficiency is the actual thrust as percentage of this.

If e.g. solar power is used this restricts $a$; in the case of a large $v_e$ the possible acceleration is inversely proportional to it, hence the time to reach a required delta-v is proportional to $v_e$; with 100% efficiency:

- for $\Delta v \ll v_e$ we have $t \approx \dfrac{mv_e\Delta v}{2P}$

Examples:

- power 1000 W, mass 100 kg, $\Delta v = 5$ km/s, $v_e = 16$ km/s, takes 1.5 months.

- power 1000 W, mass 100 kg, $\Delta v = 5$ km/s, $v_e = 50$ km/s, takes 5 months.

Thus $v_e$ should not be too large.

## Power to Thrust Ratio

The power to thrust ratio is simply:

$$\frac{P}{F} = \frac{\frac{1}{2}\dot{m}v^2}{\dot{m}v} = \frac{1}{2}v$$

Thus for any vehicle power P, the thrust that may be provided is:

$$F = \frac{P}{\frac{1}{2}v} = \frac{2P}{v}$$

## Example

Suppose a 10,000 kg space probe will be sent to Mars. The required from LEO is approximately 3000 m/s, using a Hohmann transfer orbit. For the sake of argument, assume the following thrusters are options to be used:

| Engine | Effective exhaust velocity (km/s) | Specific impulse (s) | Fuel mass (kg) | Energy required (GJ) | Energy per kg of propellant | Minimum power/ thrust | Power generator mass/ thrust |
|---|---|---|---|---|---|---|---|
| Solid rocket | 1 | 100 | 190,000 | 95 | 500 kJ | 0.5 kW/N | N/A |

| Bipropellant rocket | 5 | 500 | 8,200 | 103 | 12.6 MJ | 2.5 kW/N | N/A |
|---|---|---|---|---|---|---|---|
| Ion thruster | 50 | 5,000 | 620 | 775 | 1.25 GJ | 25 kW/N | 25 kg/N |

1.  Assuming 100% energetic efficiency; 50% is more typical in practice.

2.  Assumes a specific power of 1 kW/kg

Observe that the more fuel-efficient engines can use far less fuel; their mass is almost negligible (relative to the mass of the payload and the engine itself) for some of the engines. However, note also that these require a large total amount of energy. For Earth launch, engines require a thrust to weight ratio of more than one. To do this with the ion or more theoretical electrical drives, the engine would have to be supplied with one to several gigawatts of power, equivalent to a major metropolitan generating station. From the table it can be seen that this is clearly impractical with current power sources.

Alternative approaches include some forms of laser propulsion, where the reaction mass does not provide the energy required to accelerate it, with the energy instead being provided from an external laser or other beam-powered propulsion system. Small models of some of these concepts have flown, although the engineering problems are complex and the ground based power systems are not a solved problem.

Instead, a much smaller, less powerful generator may be included which will take much longer to generate the total energy needed. This lower power is only sufficient to accelerate a tiny amount of fuel per second, and would be insufficient for launching from Earth. However, over long periods in orbit where there is no friction, the velocity will be finally achieved. For example, it took the SMART-1 more than a year to reach the Moon, whereas with a chemical rocket it takes a few days. Because the ion drive needs much less fuel, the total launched mass is usually lower, which typically results in a lower overall cost, but the journey takes longer.

Mission planning therefore frequently involves adjusting and choosing the propulsion system so as to minimise the total cost of the project, and can involve trading off launch costs and mission duration against payload fraction.

## Rocket Engines

Most rocket engines are internal combustion heat engines (although non combusting forms exist). Rocket engines generally produce a high temperature reaction mass, as a hot gas. This is achieved by combusting a solid, liquid or gaseous fuel with an oxidiser within a combustion chamber. The extremely hot gas is then allowed to escape through a high-expansion ratio nozzle. This bell-shaped nozzle is what gives a rocket engine its characteristic shape. The effect of the nozzle is to dramatically accelerate the mass, converting most of the thermal energy into kinetic energy. Exhaust speed reaching as high as 10 times the speed of sound at sea level are common.

SpaceX's Kestrel engine is tested

Rocket engines provide essentially the highest specific powers and high specific thrusts of any engine used for spacecraft propulsion.

Ion propulsion rockets can heat a plasma or charged gas inside a magnetic bottle and release it via a magnetic nozzle, so that no solid matter need come in contact with the plasma. Of course, the machinery to do this is complex, but research into nuclear fusion has developed methods, some of which have been proposed to be used in propulsion systems, and some have been tested in a lab.

See rocket engine for a listing of various kinds of rocket engines using different heating methods, including chemical, electrical, solar, and nuclear.

## Electromagnetic Propulsion

This test engine accelerates ions using electrostatic forces

Rather than relying on high temperature and fluid dynamics to accelerate the reaction mass to high speeds, there are a variety of methods that use electrostatic or electromagnetic forces to accelerate the reaction mass directly. Usually the reaction mass is a stream of ions. Such an engine typically uses electric power, first to ionize atoms, and then to create a voltage gradient to accelerate the ions to high exhaust velocities.

The idea of electric propulsion dates back to 1906, when Robert Goddard considered the possibility in his personal notebook. Konstantin Tsiolkovsky published the idea in 1911.

For these drives, at the highest exhaust speeds, energetic efficiency and thrust are all inversely proportional to exhaust velocity. Their very high exhaust velocity means they require huge amounts of energy and thus with practical power sources provide low thrust, but use hardly any fuel.

For some missions, particularly reasonably close to the Sun, solar energy may be sufficient, and has very often been used, but for others further out or at higher power, nuclear energy is necessary; engines drawing their power from a nuclear source are called nuclear electric rockets.

With any current source of electrical power, chemical, nuclear or solar, the maximum amount of power that can be generated limits the amount of thrust that can be produced to a small value. Power generation adds significant mass to the spacecraft, and ultimately the weight of the power source limits the performance of the vehicle.

Current nuclear power generators are approximately half the weight of solar panels per watt of energy supplied, at terrestrial distances from the Sun. Chemical power generators are not used due to the far lower total available energy. Beamed power to the spacecraft shows some potential.

6 kW Hall thruster in operation at the NASA Jet Propulsion Laboratory.

Some electromagnetic methods:

- Ion thrusters (accelerate ions first and later neutralize the ion beam with an electron stream emitted from a cathode called a neutralizer)

  o Electrostatic ion thruster

  o Field-emission electric propulsion

  o Hall effect thruster

  o Colloid thruster

- Electrothermal thrusters (electromagnetic fields are used to generate a plasma to increase the heat of the bulk propellant, the thermal energy imparted to the propellant gas is then converted into kinetic energy by a nozzle of either physical material construction or by magnetic means)

- o  DC arcjet

- o  microwave arcjet

- o  Helicon Double Layer Thruster

- Electromagnetic thrusters (ions are accelerated either by the Lorentz Force or by the effect of electromagnetic fields where the electric field is not in the direction of the acceleration)

- o  Magnetoplasmadynamic thruster

- o  Electrodeless plasma thruster

- o  Pulsed inductive thruster

- o  Pulsed plasma thruster

- o  Variable specific impulse magnetoplasma rocket (VASIMR)

- Mass drivers (for propulsion)

In electrothermal and electromagnetic thrusters, both ions and electrons are accelerated simultaneously, no neutralizer is required.

## Without Internal Reaction Mass

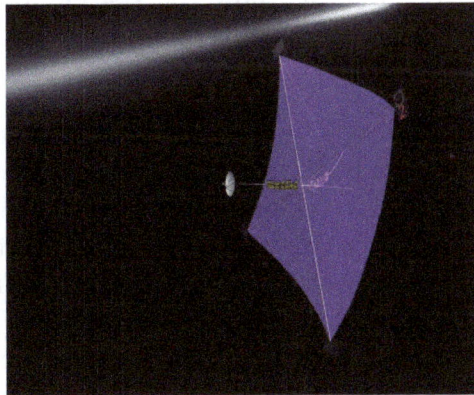

NASA study of a solar sail. The sail would be half a kilometer wide.

The law of conservation of momentum is usually taken to imply that any engine which uses no reaction mass cannot accelerate the center of mass of a spaceship (changing orientation, on the other hand, is possible). But space is not empty, especially space inside the Solar System; there are gravitation fields, magnetic fields, electromagnetic waves, solar wind and solar radiation. Electromagnetic waves in particular are known to contain momentum, despite being massless; specifically the momentum flux density P of an EM wave is quantitatively 1/c times the Poynting vector S, i.e. $P = S/c$, where c is the velocity of light. Field propulsion methods which do not rely on reaction mass thus must try to take advantage of this fact by coupling to a momentum-bearing field such as an EM wave that exists in the vicinity of the craft. However, because many of these phenomena are diffuse in nature, corresponding propulsion structures need to be proportionately large.

There are several different space drives that need little or no reaction mass to function. A tether propulsion system employs a long cable with a high tensile strength to change a spacecraft's orbit, such as by interaction with a planet's magnetic field or through momentum exchange with another object. Solar sails rely on radiation pressure from electromagnetic energy, but they require a large collection surface to function effectively. The magnetic sail deflects charged particles from the solar wind with a magnetic field, thereby imparting momentum to the spacecraft. A variant is the mini-magnetospheric plasma propulsion system, which uses a small cloud of plasma held in a magnetic field to deflect the Sun's charged particles. An E-sail would use very thin and lightweight wires holding an electric charge to deflect these particles, and may have more controllable directionality.

As a proof of concept, NanoSail-D became the first nanosatellite to orbit Earth. There are plans to add them to future Earth orbit satellites, enabling them to de-orbit and burn up once they are no longer needed. Cubesail will be the first mission to demonstrate solar sailing in low Earth orbit, and the first mission to demonstrate full three-axis attitude control of a solar sail.

Japan also launched its own solar sail powered spacecraft IKAROS in May 2010. IKAROS successfully demonstrated propulsion and guidance and is still flying today.

A satellite or other space vehicle is subject to the law of conservation of angular momentum, which constrains a body from a net change in angular velocity. Thus, for a vehicle to change its relative orientation without expending reaction mass, another part of the vehicle may rotate in the opposite direction. Non-conservative external forces, primarily gravitational and atmospheric, can contribute up to several degrees per day to angular momentum, so secondary systems are designed to "bleed off" undesired rotational energies built up over time. Accordingly, many spacecraft utilize reaction wheels or control moment gyroscopes to control orientation in space.

A gravitational slingshot can carry a space probe onward to other destinations without the expense of reaction mass. By harnessing the gravitational energy of other celestial objects, the spacecraft can pick up kinetic energy. However, even more energy can be obtained from the gravity assist if rockets are used.

## Planetary and Atmospheric Propulsion

A successful proof of concept Lightcraft test, a subset of beam-powered propulsion.

## Launch-assist Mechanisms

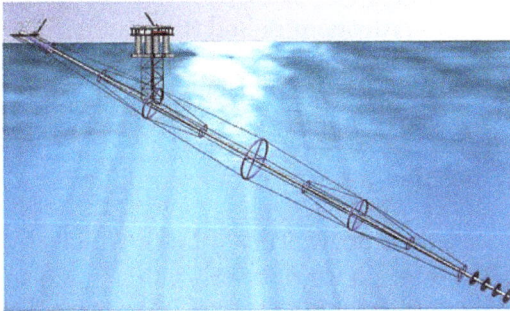

The conceptual ocean-located Quicklauncher, a light-gas gun–based space gun

There have been many ideas proposed for launch-assist mechanisms that have the potential of drastically reducing the cost of getting into orbit. Proposed non-rocket space-launch launch-assist mechanisms include:

- Skyhook (requires reusable suborbital launch vehicle, not engineeringly feasible using presently available materials)

- Space elevator (tether from Earth's surface to geostationary orbit, cannot be built with existing materials)

- Launch loop (a very fast enclosed rotating loop about 80 km tall)

- Space fountain (a very tall building held up by a stream of masses fired from its base)

- Orbital ring (a ring around Earth with spokes hanging down off bearings)

- Electromagnetic catapult (railgun, coilgun) (an electric gun)

- Rocket sled launch

- Space gun (Project HARP, ram accelerator) (a chemically powered gun)

- Beam-powered propulsion rockets and jets powered from the ground via a beam

- High-altitude platforms to assist initial stage

## Airbreathing Engines

Studies generally show that conventional air-breathing engines, such as ramjets or tur-bojets are basically too heavy (have too low a thrust/weight ratio) to give any significant performance improvement when installed on a launch vehicle itself. However, launch vehicles can be air launched from separate lift vehicles (e.g. B-29, Pegasus Rocket and White Knight) which do use such propulsion systems. Jet engines mounted on a launch rail could also be so used.

On the other hand, very lightweight or very high speed engines have been proposed that take advantage of the air during ascent:

- SABRE - a lightweight hydrogen fuelled turbojet with precooler

- ATREX - a lightweight hydrogen fuelled turbojet with precooler

- Liquid air cycle engine - a hydrogen fuelled jet engine that liquifies the air before burning it in a rocket engine

- Scramjet - jet engines that use supersonic combustion

Normal rocket launch vehicles fly almost vertically before rolling over at an altitude of some tens of kilometers before burning sideways for orbit; this initial vertical climb wastes propellant but is optimal as it greatly reduces airdrag. Airbreathing engines burn propellant much more efficiently and this would permit a far flatter launch trajectory, the vehicles would typically fly approximately tangentially to Earth's surface until leaving the atmosphere then perform a rocket burn to bridge the final delta-v to orbital velocity.

## Planetary Arrival and Landing

A test version of the MARS Pathfinder airbag system

When a vehicle is to enter orbit around its destination planet, or when it is to land, it must adjust its velocity. This can be done using all the methods listed above (provided they can generate a high enough thrust), but there are a few methods that can take advantage of planetary atmospheres and/or surfaces.

- Aerobraking allows a spacecraft to reduce the high point of an elliptical orbit by repeated brushes with the atmosphere at the low point of the orbit. This can save a considerable amount of fuel because it takes much less delta-V to enter an elliptical orbit compared to a low circular orbit. Because the braking is done over the course of many orbits, heating is comparatively minor, and a heat shield is not required. This has been done on several Mars missions such as Mars Global Surveyor, Mars Odyssey and Mars Reconnaissance Orbiter, and at least one Venus mission, Magellan.

- Aerocapture is a much more aggressive manoeuver, converting an incoming hyperbolic orbit to an elliptical orbit in one pass. This requires a heat shield and much trickier navigation, because it must be completed in one pass through the atmosphere, and unlike aerobraking no preview of the atmosphere is possible. If the intent is to remain in orbit, then at least one more propulsive maneuver is required after aerocapture—otherwise the low point of the resulting orbit will remain in the atmosphere, resulting in eventual re-entry. Aerocapture has not yet been tried on a planetary mission, but the re-entry skip by Zond 6 and Zond 7 upon lunar return were aerocapture maneuvers, because they turned a hyperbolic orbit into an elliptical orbit. On these missions, because there was no attempt to raise the perigee after the aerocapture, the resulting orbit still intersected the atmosphere, and re-entry occurred at the next perigee.

- A ballute is an inflatable drag device.

- Parachutes can land a probe on a planet or moon with an atmosphere, usually after the atmosphere has scrubbed off most of the velocity, using a heat shield.

- Airbags can soften the final landing.

- Lithobraking, or stopping by impacting the surface, is usually done by accident. However, it may be done deliberately with the probe expected to survive, in which case very sturdy probes are required.

## Hypothetical Methods

A variety of hypothetical propulsion techniques have been considered that would require entirely new principles of physics to be realized or that may not exist. To date, such methods are highly speculative and include:

- Diametric drive

- Pitch drive & bias drive

- Disjunction drive

- Alcubierre drive (a form of warp drive)

- Differential sail

- Wormholes – theoretically possible, but unachieveable in practice with current technology

- Woodward effect

- Reactionless drives – breaks the law of conservation of momentum; theoretically impossible

- Photon rocket

- Bussard ramjet

- A "hyperspace" drive based upon Heim theory

- Micronewton electromagnetic thruster - Linear momentum loss has been claimed for an electromagnetically powered thruster

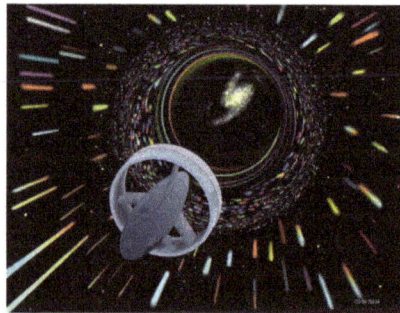

Artist's conception of a warp drive design

A NASA assessment is found at Marc G Millis *Assessing potential propulsion breakthroughs* (2005) and an overview of NASA research in this area is at Breakthrough Propulsion Physics.

## References

- King-Hele, Desmond (1987). Satellite orbits in an atmosphere: Theory and application. Springer. ISBN 978-0-216-92252-5.

- Dimitri S.H. Charrier (2012). "Micronewton electromagnetic thruster". Applied Physics Letters. 101. p. 034104.

# 3

# Various Satellite Navigation Systems

There are several satellite navigation systems in use globally. This chapter provides the reader with elaborate material on different satellite navigation systems such as GPS Aided GEO Augmented Navigation, TRANSIT system, BeiDou Navigation Satellite System and Galileo. The content describes the main objectives, international involvement, system description and impact of these systems on modern technology.

## Transit (Satellite)

Transit 2A with GRAB 1 atop during launch preparations

The TRANSIT system, also known as NAVSAT or NNSS (for *Navy Navigation Satellite System*), was the first satellite navigation system to be used operationally. The system was primarily used by the U.S. Navy to provide accurate location information to its Polaris ballistic missile submarines, and it was also used as a navigation system by the Navy's surface ships, as well as for hydrographic survey and geodetic surveying. Transit provided continuous navigation satellite service from 1964, initially for Polaris submarines and later for civilian use as well.

### History

The TRANSIT satellite system sponsored by the Navy and developed jointly by DARPA and the Johns Hopkins Applied Physics Laboratory, under the leadership of Dr. Richard Kirschner at Johns Hopkins was the first satellite positioning system. Just days after the Soviet launch of Sputnik 1, the first man-made earth-orbiting satellite on October 4, 1957, two physicists at APL, William Guier and George Weiffenbach, found themselves in discussion about the microwave signals that would likely be emanating from the satellite. They were able to determine Sputnik's orbit by analyzing the Doppler shift of its radio signals during a single pass. Discussing the way forward for their research, their director Frank McClure, the chairman of APL's Research Center, suggested in March 1958 that if the satellite's position were known and predictable, the Doppler shift could be used to locate a receiver on Earth, and proposed a satellite system to implement this principle.

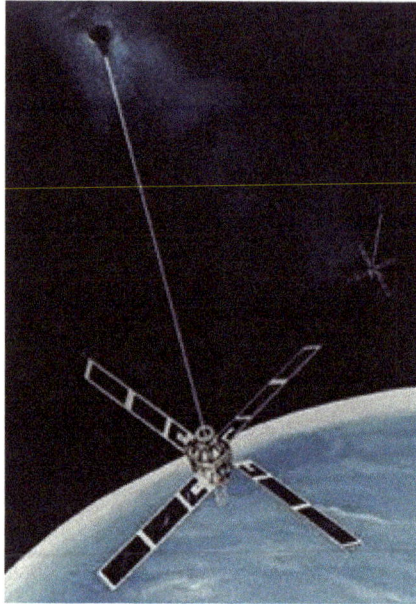

Operational Transit satellite

Development of the TRANSIT system began in 1958, and a prototype satellite, Transit 1A, was launched in September 1959. That satellite failed to reach orbit. A second satellite, Transit 1B, was successfully launched April 13, 1960, by a Thor-Ablestar rocket. The first successful tests of the system were made in 1960, and the system entered Naval service in 1964.

The Chance Vought/LTV Scout rocket was selected as the dedicated launch vehicle for the program because it delivered a payload into orbit for the lowest cost per pound. However, the Scout decision imposed two design constraints. First, the weights of the earlier satellites were about 300 lb each, but the Scout launch capacity to the Transit orbit was about 120 lb (it was later increased significantly). A satellite mass reduction had to be achieved despite a demand for more power than APL had previously designed into a satellite. The second problem concerned the increased vibration that affected the payload during launching because the Scout used solid rocket motors. Thus, electronic equipment that was smaller than before and rugged enough to withstand the increased vibration of launch had to be produced. Meeting the new demands was more difficult than expected, but it was accomplished. The first prototype operational satellite

(Transit 5A-1) was launched into a polar orbit by a Scout rocket on 18 December 1962. The satellite verified a new technique for deploying the solar panels and for separating from the rocket, but otherwise it was not successful because of trouble with the power system. Transit 5A-2, launched on 5 April 1963, failed to achieve orbit. Transit 5A-3, with a redesigned power supply, was launched on 15 June 1963. A malfunction of the memory occurred during powered flight that kept it from accepting and storing the navigation message, and the oscillator stability was degraded during launch. Thus, 5A-3 could not be used for navigation. However, this satellite was the first to achieve gravity-gradient stabilization, and its other subsystems performed well.

Surveyors used Transit to locate remote benchmarks by averaging dozens of Transit fixes, producing sub-meter accuracy. In fact, the elevation of Mount Everest was corrected in the late 1980s by using a Transit receiver to re-survey a nearby benchmark.

Thousands of warships, freighters and private watercraft used Transit from 1967 until 1991. In the 1970s, the Soviet Union started launching their own satellite navigation system *Parus* (military) / Tsikada (civilian), that is still in use today besides the next generation GLONASS. Some Soviet warships were equipped with Motorola NavSat receivers.

The TRANSIT system was made obsolete by the Global Positioning System (GPS), and ceased navigation service in 1996. Improvements in electronics allowed the GPS system to effectively take several fixes at once, greatly reducing the complexity of deducing a position. The GPS system uses many more satellites than were used with TRANSIT, allowing the system to be used continuously, while TRANSIT provided a fix only every hour or more.

After 1996, the satellites were kept in use for the Navy Ionospheric Monitoring System (NIMS).

## Description

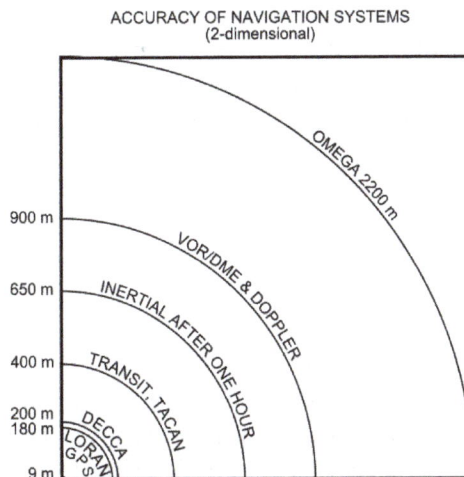

ACCURACY OF NAVIGATION SYSTEMS
(2-dimensional)

## Satellites

The satellites (known as *OSCAR* or *NOVA* satellites) used in the system were placed in low polar orbits, at an altitude of about 600 nautical miles (1,100 km), with an orbital period of about 106 minutes. A *constellation* of five satellites was required to provide reasonable global coverage. While the system was operational, at least ten satellites – one spare for each satellite in the basic constellation – were usually kept in orbit. Note that these *OSCAR* satellites were not the same as the OSCAR series of satellites that were devoted to use by amateur radio operators to use in satellite communications.

Transit-1-Satellite Prototype

The orbits of the TRANSIT satellites were chosen to cover the entire Earth; their orbits crossed over the poles and were spread out at the equator. Since only one was usually visible at any given time, fixes could be made only when one of the satellites was above the horizon. At the equator this delay between fixes was several hours; at mid-latitudes the delay decreased to an hour or two. For its intended role as an updating system for SLBM launch, TRANSIT sufficed, since submarines took periodic fixes to re-set their inertial guidance system, but TRANSIT lacked the ability to provide high-speed, real-time position measurements.

With later improvements, the system provided single-pass accuracy of roughly 200 meters, and also provided time synchronization to roughly 50 microseconds. TRANSIT satellites also broadcast encrypted messages, although this was a secondary function.

The Transit satellites used arrays of magnetic-core memory as mass data storage up to 32 kilobytes.

## Determining Ground Location

The basic operating principle of TRANSIT is similar to the system used by emergency locator transmitters, except their transmitter is on the ground and the receiver is in orbit. Details on the signal are forwarded directly to ground stations, which then generate a fix on the transmitter using a process similar to TRANSIT.

The TRANSIT system satellites broadcast two UHF carrier signals that provided precise time hacks (every two minutes), plus the satellite's six orbital elements and orbit perturbation variables. The orbit ephemeris and clock corrections were uploaded twice each day to each satellite from one of the four Navy tracking and injection stations. This broadcast information allowed a ground receiver to calculate the location of the satellite at any point in time. Use of two carriers permitted ground receivers to reduce navigation errors caused by ionospheric refraction. The Transit system also provided the first world-wide time-keeping service, allowing clocks everywhere to be synchronised with 50 micro-second accuracy.

The transit satellite broadcast on 150 and 400 MHz. The two frequencies were used to allow the bending of the satellite radio beacons by the ionosphere to be canceled out, thereby improving location accuracy.

The critical information that allowed the receiver to compute location was a unique frequency curve caused by the Doppler effect. The Doppler effect caused an apparent compression of the carrier's wavelength as the satellite approached the receiver, and stretching of wavelengths as the satellite receded. The spacecraft traveled at about 17,000 mph, which could increase or decrease the received carrier signal by as much as 10 kHz. This Doppler curve was unique for each location within line-of-sight of the satellite. For instance, the earth's rotation caused the ground receiver to move toward or away from the satellite's orbit, creating a non-symmetric Doppler shift for approach and recession, allowing the receiver to determine whether it was east or west of the satellite's north-south ground track.

Calculating the most likely receiver location was not a trivial exercise. The navigation software used the satellite's motion to compute a 'trial' Doppler curve, based on an initial 'trial' location for the receiver. The software would then perform a least squares curve fit for each two-minute section of the Doppler curve, recursively moving the trial position until the trial Doppler curve 'most closely' matched the actual Doppler received from the satellite for all 2-minute curve segments.

If the receiver was also moving relative to the earth, such as aboard a ship or airplane, this would cause mismatches with the idealized Doppler curves, and degrade position accuracy. However, positional accuracy could usually be computed to within 100-meters for a slow-moving ship, even with reception of just one two-minute Doppler curve. This was the navigation criterion demanded by the U.S. Navy, since American submarines would normally expose their UHF antenna for only 2-minutes to obtain a usable Transit fix. The U.S. Submarine version of the Transit System also included a special encrypted (and more accurate) version of the downloaded satellite's orbital data. This enhanced data allowed for considerably enhanced system accuracy [not unlike Selective Availability (SA) under GPS]. Using this enhanced mode accuracy was typically less than 20 meters. (Between LORAN C and GPS.) Certainly, the most accurate navigation system of its day.

## Determining the Satellite Orbits

Vestibule and Quonset hut housing Transit satellite tracking station 019. 1. Triad satellite magnetometer down load antenna. 2. flag pole, 3. Utility pole in background, 4 Revolving light temperature alarm, 5 VLF antenna, 6-9 Doppler satellite tracking antennas, 10. stove pipe for heater, 11 Flood light for low visibility conditions, 12 fuel tank.

A network of ground stations, whose locations were accurately known, continually tracked the transit satellites. They measured the Doppler shift and transferred the data to 5 hole paper tape using standard teleprinter hole convention. This data was sent to the Satellite Control Center at Applied Physics Laboratory in Laurel, Maryland using commercial and military teleprinter networks. The data from the fixed ground stations provided the location information on the transit satellite orbit. Locating a transit satellite in earth orbit from a known ground station using the Doppler shift is simply the reverse of using the known location of the satellite in orbit to locate an unknown location on the earth, again using the Doppler shift .

Some of the equipment inside Transit satellite tracking station 019. 1. Automatic Control Unit, 2. timer-counter, 3. Time burst detector, 4. time conversion chart, 5. satellite ephemeris, 6. tracking receiver, 7. time display, 8 Header-Tailer programmer, 9. Digitizer and main clock, 10. master oscillator, 11. strip chart recorder, 12. paper tape punch, 13. short wave receiver. Out of sight: VLF receiver, refraction correction unit, backup battery system, power supplies, AC voltage regulators.

A typical ground station occupied a small Quonset hut. The accuracy of the ground station measurements was a function of the ground station master clock accuracy. Initially a quartz oscillator in a temperature controlled oven was used as the master clock. The master clock was checked daily for drift using a VLF receiver tuned to a US Navy low frequency VLF station. The VLF signal had the property that the phase of the VLF signal did not change from day to day at noon along the path between the transmitter and the receiver and thus could be used to measure oscillator drift. Later rubidium beam and cesium beam clocks were used. Ground stations had number names; for example, Station 019 was McMurdo Station, Antarctica. For many years during the 1970s this station was staffed by a graduate student and an undergraduate student, typically in electrical engineering, from the University of Texas at Austin. Other stations were located at New Mexico State University, the University of Texas at Austin, Sicily, Japan, Seychelles Island, Thule Greenland and a number of other locations. The Greenland and Antarctica stations saw every pass of every transit satellite because of their near pole location for these polar orbiting satellites.

## Portable Geoceiver

A portable version of the ground station was called a Geoceiver and was used to make field measurements. This receiver, power supply, punched tape unit, and antennas could fit in a number of padded aluminum cases and could be shipped as extra cargo on an airline. Data was taken over a period of time, typically a week, and sent back to the Satellite Control Center for processing. Therefore, unlike GPS, there was not an immediate accurate location of the geoceiver location. A geoceiver was permanently located at the South Pole Station and operated by USGS personnel. Since it was located on the surface of a moving ice sheet, its data was used to measure the ice sheet movement. Other geoceivers were taken out in the field in Antarctica during the summer and were used to measure locations, for example the movement of the Ross Ice Shelf.

## The AN/UYK-1 Computer

Since no computer small enough to fit through a submarine's hatch existed (in 1958), a new computer was designed, named the AN/UYK-1. It was built with rounded corners to fit through the hatch and was about five feet tall and sealed to be waterproof. The principal design engineer was then-UCLA-faculty-member Lowell Amdahl, brother of Gene Amdahl. The AN/UYK-1 was built by the Ramo-Wooldridge Corporation (later TRW) for the *Lafayette* class SSBNs. It was equipped with 8,192 words of 15-bit core memory plus parity bit, threaded by hand at their Canoga Park factory. Cycle time was about one microsecond.

The AN/UYK-1 was a "micro-programmed" machine with a 15-bit word length that lacked hardware commands to subtract, multiply or divide, but could add, shift, form one's complement, and test the carry bit. Instructions to perform standard fixed and floating point operations were software subroutines and programs were lists of links and operators to those subroutines. For example, the "subtract" subroutine had to form the one's complement of the subtrahend and add it. Multiplication required successive shifting and conditional adding.

The most interesting feature of the AN/UYK-1 instruction set was that the machine-language instructions had two operators that could simultaneously manipulate the arithmetic registers, for example complementing the contents of one register while loading or storing another. It also may have been the first computer that implemented a single-cycle indirect addressing ability.

During a satellite pass, a GE receiver would receive the orbital parameters and encrypted messages from the satellite, as well as measure the Doppler shifted frequency at intervals and provide this data to the AN/UYK-1 computer. The computer would also receive from the ship's inertial navigation system (SINS) a reading of latitude and longitude. Using this information the AN/UYK-1 ran the least squares algorithm and provided a location reading in about fifteen minutes.

## Other Satellites

There were 37 other satellites in the Transit series that were assigned the Transit name by NASA.

Transit 3B demonstrated uploading programs into the onboard computer's memory whilst in orbit.

Transit 4A, launched June 29, 1961, was the first satellite to use a radioactive power source (a SNAP-3). Transit 4B (1961) also had a SNAP-3 RTG. Transit 4B was among several satellites which were inadvertently damaged or destroyed in a nuclear explosion, specifically the United States Starfish Prime high-altitude nuclear test on July 9, 1962 and subsequent radiation belt.

Transit 5A3 and Transit 5B-1 (1963) each had a SNAP-3 RTG.

Transit 5B-2 (1963) had a SNAP-9A RTG.

Transit-9 and 5B4 (1964) and Transit-5B7 and 5B6 (1965) each had "a nuclear power source".

The US Air Force also periodically launched short lived satellites equipped with radio beacons of 162 MHz and 324 MHz at much lower orbits to study orbital drag. The Transit ground tracking stations tracked these satellites as well, locating the satellites within their orbits using the same principles. The satellite location data was used to collect orbital drag data, including variations in the upper atmosphere and the Earth's gravitational field.

# GPS Aided GEO Augmented Navigation

The GPS Aided GEO Augmented Navigation (GAGAN) is an implementation of a regional satellite-based augmentation system (SBAS) by the Indian government. It is a system to improve the accuracy of a GNSS receiver by providing reference signals.

The AAI's efforts towards implementation of operational SBAS can be viewed as the first step towards introduction of modern communication, navigation, surveillance/Air Traffic Management system over Indian airspace.

The project has established 15 Indian Reference Stations, 3 Indian Navigation Land Uplink Stations, 3 Indian Mission Control Centers, and installation of all associated software and communication links. It will be able to help pilots to navigate in the Indian airspace by an accuracy of 3 m. This will be helpful for landing aircraft in tough weather and terrain like Mangalore and Leh airports.

## Implementation

The ₹7.74 billion (US$115 million) project is being created in three phases through 2008 by the Airport Authority of India with the help of the Indian Space Research Organization's (ISRO) technology and space support. The goal is to provide navigation system for all phases of flight over the Indian airspace and in the adjoining area. It is applicable to safety-to-life operations, and meets the performance requirements of international civil aviation regulatory bodies.

The space component will become available after the GAGAN payload on the GSAT-8 communication satellite, which was launched recently, is switched on. This payload was also on the GSAT-4 satellite that was lost when the Geosynchronous Satellite Launch Vehicle (GSLV) failed during launch in April 2010. Final System Acceptance Test was conducted during June 2012 followed by system certification during July 2013.

## Technology

To begin implementing a satellite-based augmentation system over the Indian airspace, Wide Area Augmentation System (WAAS) codes for L1 frequency and L5 frequency were obtained from the United States Air Force and U.S Department of Defense on November 2001 and March 2005. The system will use eight reference stations located in Delhi, Guwahati, Kolkata, Ahmedabad, Thiruvananthapuram, Bangalore, Jammu and Port Blair, and a master control center at Bangalore. US defense contractor Raytheon has stated they will bid to build the system.

## Technology Demonstration

A national plan for satellite navigation including implementation of Technology Demonstration System (TDS) over the Indian air space as a proof of concept had been prepared jointly by Airports Authority of India (AAI) and ISRO. TDS was successfully completed during 2007 by installing eight Indian Reference Stations (INRESs) at eight Indian airports and linked to the Master Control Center (MCC) located near Bangalore. Preliminary System Acceptance Testing has been successfully completed in December 2010. The ground segment for GAGAN, which has been put up by the Raytheon, has 15 reference stations scattered across the country. Two mission control centres, along with associated uplink stations, have been set up at Kundalahalli in Bangalore. One more control centre and uplink station are to come up at Delhi. As a part of the programme, a network of 18 total electron content (TEC) monitoring stations were installed at various locations in India to study and analyse the behaviour of the ionosphere over the Indian region.

GAGAN's TDS signal in space provides a three-metre accuracy as against the requirement of 7.6 metres. Flight inspection of GAGAN signal is being carried out at Kozhikode, Hyderabad, Nagpur and Bangalore airports and the results have been satisfactory so far.

## Study of Ionosphere

One essential component of the GAGAN project is the study of the ionospheric behavior over the Indian region. This has been specially taken up in view of the rather uncertain nature of the behavior of the ionosphere in the region. The study will lead to the optimization of the algorithms for the ionospheric corrections in the region.

To study the ionospheric behavior more effectively over entire Indian airspace, Indian universities and R&D labs, which are involved in the development of regional based ionotropic model for GAGAN, have suggested nine more TEC stations.

## Technology Integration

GAGAN after its final operational phase completion, will be compatible with other SBAS systems such as the Wide Area Augmentation System (WAAS), the European Geostationary Navigation Overlay Service (EGNOS) and the Multi-functional Satellite Augmentation System (MSAS) and will provide seamless air navigation service across regional boundaries. While the ground segment consists of eight reference stations and a master control center, which will have sub systems such as data communication network, SBAS correction and verification system, operations and maintenance system, performance monitoring display and payload simulator, Indian land uplinking stations will have dish antenna assembly. The space segment will consist of one geo-navigation transponder.

## Effective Flight-management System

A flight-management system based on GAGAN will then be poised to save operators time and money by managing climb, descent and engine performance profiles. The FMS will improve the efficiency and flexibility by increasing the use of operator-preferred trajectories. It will improve airport and airspace access in all weather conditions, and the ability to meet the environmental and obstacle clearance constraints. It will also enhance reliability and reduce delays by defining more precise terminal area procedures that feature parallel routes and environmentally optimised airspace corridors.

- GAGAN will increase safety by using a three-dimensional approach operation with course guidance to the runway, which will reduce the risk of controlled flight into terrain i.e., an accident whereby an airworthy aircraft, under pilot control, inadvertently flies into terrain, an obstacle, or water.

- GAGAN will also offer high position accuracies over a wide geographical area like the Indian airspace. These positions accuracies will be simultaneously available to 80 civilian and more than 200 non-civilian airports and airfields and will facilitate an increase in the number of airports to 500 as planned. These position accuracies can be further enhanced with ground based augmentation system.

## Developments

The first GAGAN transmitter was integrated into the GSAT-4 geostationary satellite, and had a goal of being operational in 2008. Following a series of delays, GSAT-4 was launched on 15 April 2010, however it failed to reach orbit after the third stage of the Geosynchronous Satellite Launch Vehicle Mk.II that was carrying it malfunctioned.

In 2009, Raytheon had won an 82 million dollar contract. It was mainly dedicated to modernize Indian air navigation system. The vice president of Command & Control Systems, Raytheon Network Centric Systems, Andy Zogg commented:

"GAGAN will be the world's most advanced air navigation system and further reinforces India's leadership in the forefront of air navigation. GAGAN will greatly improve safety, reduce congestion and enhance communications to meet India's growing air traffic management needs"

In 2012, the Defence Research and Development Organisation received a "miniaturised version" of the device with all the features from global positioning systems (GPS) and global navigation satellite systems (GNSS). The module weighing just 17 gm, can be used in multiple platforms ranging from aircraft (e.g. winged or rotor-craft) to small boats, ships. Reportedly, it can also assist "survey applications". It is a cost-efficient device and can be of "tremendous" civilian use. The navigation output is composed of GPS, GLONASS and GPS+GLONASS position, speed and time data. According to a statement released by the DRDO, G30M is a state-of-the-art technology receiver, integrating Indian GAGAN as well as both global positioning system and GLONASS systems.

According to Deccan chronicle:

"G. Satheesh Reddy, associate director of the city-based Research Centre Imarat, said the product is bringing about a quantum leap in the area of GNSS technology and has paved the way for highly miniaturised GNSS systems for the future."

On 30 December 2013, the Directorate General of Civil Aviation (DGCA), India provisionally certified the GPS Aided Geo Augmented Navigation (GAGAN) system to RNP0.1 (Required Navigation Performance, 0.1 Nautical Mile) service level. The certification enabled aircraft fitted with SBAS equipment to use GAGAN signal in space for navigation purposes.

## Satellites

GSAT-8 is an Indian geostationary satellites, which was successfully launched using Ariane 5 on 21 May 2011 and is positioned in geosynchronous orbit at 55 degrees E longitude.

GSAT-10 is envisaged to augment the growing need of Ku and C-band transponders and carries 12 Ku Band, 12 C Band and 12 Extended C Band transponders and a GAGAN payload. The spacecraft employs the standard I-3K structure with power handling capability of around 6 kW with a lift off mass of 3400 kg. GSAT-10 was successfully launched by Ariane 5 on 29 September 2012.

GSAT-15 carries 24 Ku band transponders with India coverage beam and a GAGAN payload. was successfully launched on 10 November 2015, 21:34:07 UTC, completing the constellation.

## Indian Regional Navigation Satellite System

The Indian government has stated that it intends to use the experience of creating the GAGAN system to enable the creation of an autonomous regional navigation system called the Indian Regional Navigation Satellite System IRNSS.

IRNSS-1 Indian Regional Navigational Satellite System (IRNSS)-1, the first of the seven satellites of the IRNSS constellation, carries a Navigation payload and a C-band ranging transponder. The spacecraft employs an optimized I-1K structure with a power handling capability of around 1660W and a lift off mass of 1425 kg, and is designed for a nominal mission life of 10 years. The first satellite of IRNSS constellation was launched onboard PSLV (C22) on 1 July 2013. While the full constellation was planned to be realized during 2014 time frame, launch of subsequent satellites got delayed. Currently all 7 satellites are in orbit, and the system is expected to start operating by June 2016.

## Applications

Karnataka Forest Department has used GAGAN to build a new, accurate and publicly available satellite based database of its forestlands. This is a followup to the Supreme Court directive to States to update and put up their respective forest maps. The geospatial database of forestlands pilot has used data from the Cartosat-2 satellite. The maps are meant to rid authorities of ambiguities related to forest boundaries and give clarity to forest administrators, revenue officials as also the public, according to R.K. Srivastava, Chief Conservator of Forests (Headquarters).

Various Indian manufactured missiles including the BrahMos will use GAGAN for guidance.

# BeiDou Navigation Satellite System

The BeiDou Navigation Satellite System (BDS, simplified Chinese: traditional Chinese: pinyin: *Běidǒu wèixīng dǎoháng xìtǒng*) is a Chinese satellite navigation system. It consists of two separate satellite constellations – a limited test system that has been operating since 2000, and a full-scale global navigation system that is currently under construction.

The first BeiDou system, officially called the BeiDou Satellite Navigation Experimental System (simplified Chinese: traditional Chinese: pinyin: *Běidǒu wèixīng dǎoháng shìyàn xìtǒng*) and also known as BeiDou-1, consists of three satellites and offers limited coverage and applications. It has been offering navigation services, mainly for customers in China and neighboring regions, since 2000.

The second generation of the system, officially called the BeiDou Navigation Satellite System (BDS) and also known as COMPASS or BeiDou-2, will be a global satellite navigation system consisting of 35 satellites, and is under construction as of January 2015. It became operational in China in December 2011, with 10 satellites in use, and began offering services to customers in the Asia-Pacific region in December 2012. It is planned to begin serving global customers upon its completion in 2020.

In-mid 2015, China started the build-up of the third generation BeiDou system (BDS-3) in the global coverage constellation. The first BDS-3 satellite was launched 30 September 2015. As of March 2016, 4 BDS-3 in-orbit validation satellites have been launched.

According to China Daily, fifteen years after the satellite system was launched, it is now generating a turnover of $31.5 billion per annum for major companies such as China Aerospace Science and Industry Corp, AutoNavi Holdings Ltd, and China North Industries Group Corp.

## Nomenclature

The official English name of the system is *BeiDou Navigation Satellite System*. It is named after the Big Dipper constellation, which is known in Chinese as *Běidǒu*. The name literally means "Northern Dipper", the name given by ancient Chinese astronomers to the seven brightest stars of the Ursa Major constellation. Historically, this set of stars was used in navigation to locate the North Star Polaris. As such, the name BeiDou also serves as a metaphor for the purpose of the satellite navigation system.

## History

## Conception and Initial Development

The original idea of a Chinese satellite navigation system was conceived by Chen Fangyun and his colleagues in the 1980s. According to the China National Space Administration, the development of the system would be carried out in three steps:

1. 2000–2003: experimental BeiDou navigation system consisting of 3 satellites

2. by 2012: regional BeiDou navigation system covering China and neighboring regions

3. by 2020: global BeiDou navigation system

The first satellite, *BeiDou-1A*, was launched on 30 October 2000, followed by *BeiDou-1B* on 20 December 2000. The third satellite, *BeiDou-1C* (a backup satellite), was put into orbit on 25 May 2003. The successful launch of *BeiDou-1C* also meant the establishment of the BeiDou-1 navigation system.

On 2 November 2006, China announced that from 2008 BeiDou would offer an open service with an accuracy of 10 meters, timing of 0.2 microseconds, and speed of 0.2 meters/second.

In February 2007, the fourth and last satellite of the BeiDou-1 system, *BeiDou-1D* (sometimes called *BeiDou-2A*, serving as a backup satellite), was sent up into space. It was reported that the satellite had suffered from a control system malfunction but was then fully restored.

In April 2007, the first satellite of BeiDou-2, namely *Compass-M1* (to validate frequencies for the BeiDou-2 constellation) was successfully put into its working orbit. The second BeiDou-2 constellation satellite *Compass-G2* was launched on 15 April 2009. On 15 January 2010, the official website of the BeiDou Navigation Satellite System went online, and the system's third satellite (*Compass-G1*) was carried into its orbit by a Long March 3C rocket on 17 January 2010. On 2 June 2010, the fourth satellite was launched successfully into orbit. The fifth orbiter was launched into space from Xichang Satellite Launch Center by an LM-3I carrier rocket on 1 August 2010. Three months later, on 1 November 2010, the sixth satellite was sent into orbit by LM-3C. Another satellite, the Beidou-2/Compass IGSO-5 (fifth inclined geosynchonous orbit) satellite, was launched from the Xichang Satellite Launch Center by a Long March-3A on 1 December 2011 (UTC).

## Chinese Involvement in Galileo System

In September 2003, China intended to join the European Galileo positioning system project and was to invest €230 million (USD296 million, GBP160 million) in Galileo over the next few years. At the time, it was believed that China's "BeiDou" navigation system would then only be used by its armed forces. In October 2004, China officially joined the Galileo project by signing the *Agreement on the Cooperation in the Galileo Program between the "Galileo Joint Undertaking" (GJU) and the "National Remote Sensing Centre of China" (NRSCC)*. Based on the Sino-European Cooperation Agreement on Galileo program, China Galileo Industries (CGI), the prime contractor of the China's involvement in Galileo programs, was founded in December 2004. By April 2006, eleven cooperation projects within the Galileo framework had been signed between China and EU. However, the Hong Kong-based *South China Morning Post* reported in January 2008 that China was unsatisfied with its role in the Galileo project and was to compete with Galileo in the Asian market.

## Phase III

- In November 2014, Beidou became part of the World-Wide Radionavigation System (WWRNS) at the 94th meeting of The International Maritime Organization (IMO) Maritime Safety Committee, which approved the "Navigation Safety Circular" of the Beidou Navigation Satellite System (BDS).

- At Beijing time 21:52, March 30, 2015, the first new-generation BeiDou Navigation satellite (and the 17th overall) was successfully set to orbit by a Long March 3C rocket.

## Experimental System (BeiDou-1)

## Description

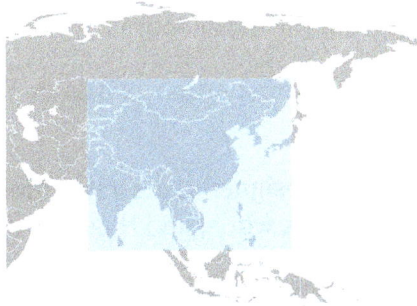

Coverage polygon of BeiDou-1.

BeiDou-1 is an experimental regional navigation system, which consists of four satellites (three working satellites and one backup satellite). The satellites themselves were based on the Chinese DFH-3 geostationary communications satellite and had a launch weight of 1,000 kilograms (2,200 pounds) each.

Unlike the American GPS, Russian GLONASS, and European Galileo systems, which use medium Earth orbit satellites, BeiDou-1 uses satellites in geostationary orbit. This means that the system does not require a large constellation of satellites, but it also limits the coverage to areas on Earth where the satellites are visible. The area that can be serviced is from longitude 70°E to 140°E and from latitude 5°N to 55°N. A frequency of the system is 2491.75 MHz.

## Completion

The first satellite, BeiDou-1A, was launched on October 31, 2000. The second satellite, BeiDou-1B, was successfully launched on December 21, 2000. The last operational satellite of the constellation, BeiDou-1C, was launched on May 25, 2003.

## Position Calculation

In 2007, the official Xinhua News Agency reported that the resolution of the BeiDou system was as high as 0.5 metres. With the existing user terminals it appears that the calibrated accuracy is 20m (100m, uncalibrated).

## Terminals

In 2008, a BeiDou-1 ground terminal cost around CN¥ 20,000RMB (US$2,929), almost 10 times the price of a contemporary GPS terminal. The price of the terminals was explained as being due to the cost of imported microchips. At the China High-Tech Fair ELEXCON of November 2009 in Shenzhen, a BeiDou terminal priced at CN¥ 3,000RMB was presented.

## Applications

- Over 1,000 BeiDou-1 terminals were used after the 2008 Sichuan earthquake, providing information from the disaster area.

- As of October 2009, all Chinese border guards in Yunnan are equipped with BeiDou-1 devices.

According to Sun Jiadong, the chief designer of the navigation system, "Many organizations have been using our system for a while, and they like it very much."

The new-generation BeiDou satellites support short message service.

## Global System (BeiDou Navigation Satellite System or BeiDou-2)
## Description

BeiDou-2 (formerly known as COMPASS) is not an extension to the older BeiDou-1, but rather supersedes it outright. The new system will be a constellation of 35 satellites, which include 5 geostationary orbit satellites for backward compatibility with BeiDou-1, and 30 non-geostationary satellites (27 in medium Earth orbit and 3 in inclined geosynchronous orbit), that will offer complete coverage of the globe.

Coverage polygon of BeiDou-2 in 2012.

The ranging signals are based on the CDMA principle and have complex structure typical of Galileo or modernized GPS. Similar to the other global navigation satellite systems (GNSSs), there will be two levels of positioning service: open (public) and restricted (military). The public service will be available globally to general users. When all the currently planned GNSSs are deployed, users of multi-constellation receivers will benefit from a total over 100 satellites, which will significantly improve all aspects of positioning, especially availability of the signals in so-called urban canyons. The general designer of the COMPASS navigation system is Sun Jiadong, who is also the general designer of its predecessor, the original BeiDou navigation system.

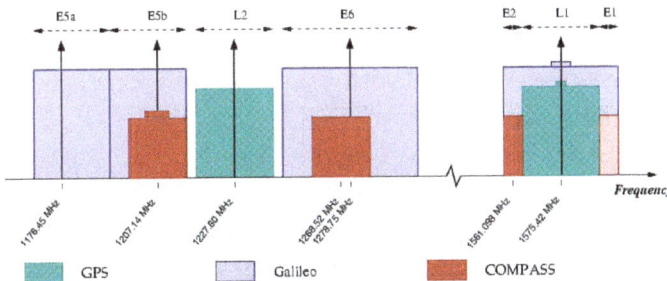

Frequency allocation of GPS, Galileo, and COMPASS; the light red color of E1 band indicates that the transmission in this band has not yet been detected.

## Accuracy

There are two levels of service provided — a free service to civilians and licensed service to the Chinese government and military. The free civilian service has a 10-meter location-tracking accuracy, synchronizes clocks with an accuracy of 10 nanoseconds, and measures speeds to within 0.2 m/s. The restricted military service has a location accuracy of 10 centimetres, can be used for communication, and will supply information about the system status to the user. To date, the military service has been granted only to the People's Liberation Army and to the Military of Pakistan.

## Frequencies

Frequencies for COMPASS are allocated in four bands: E1, E2, E5B, and E6 and overlap with Galileo. The fact of overlapping could be convenient from the point of view of the receiver design, but on the other hand raises the issues of inter-system interference, especially within E1 and E2 bands, which are allocated for Galileo's publicly regulated service. However, under International Telecommunication Union (ITU) policies, the first nation to start broadcasting in a specific frequency will have priority to that frequency, and any subsequent users will be required to obtain permission prior to using that frequency, and otherwise ensure that their broadcasts do not interfere with the original nation's broadcasts. It now appears that Chinese COMPASS satellites will start transmitting in the E1, E2, E5B, and E6 bands before Europe's Galileo satellites and thus have primary rights to these frequency ranges.

Although little was officially announced by Chinese authorities about the signals of the new system, the launch of the first COMPASS satellite permitted independent researchers not only to study general characteristics of the signals, but even to build a COMPASS receiver.

## Compass-M1

Compass-M1 is an experimental satellite launched for signal testing and validation and for the frequency filing on 14 April 2007. The role of Compass-M1 for Compass is similar to the role of the GIOVE satellites for the Galileo system. The orbit of Compass-M1 is nearly circular, has an altitude of 21,150 km and an inclination of 55.5 degrees.

Compass-M1 transmits in 3 bands: E2, E5B, and E6. In each frequency band two coherent sub-signals have been detected with a phase shift of 90 degrees (in quadrature). These signal components are further referred to as "I" and "Q". The "I" components have shorter codes and are likely to be intended for the open service. The "Q" components have much longer codes, are more interference resistive, and are probably intended for the restricted service. IQ modulation has been the method in both wired and wireless digital modulation since morsetting carrier signal 100 years ago.

The investigation of the transmitted signals started immediately after the launch of Compass -M1 on 14 April 2007. Soon after in June 2007, engineers at CNES reported the spectrum and structure of the signals. A month later, researchers from Stanford University reported the complete decoding of the "I" signals components. The knowledge of the codes allowed a group of engineers at Septentrio to build the COMPASS receiver and report tracking and multipath characteristics of the "I" signals on E2 and E5B.

| Characteristics of Compass signals reported as of May 2008 compared to GPS-L1CA | | | | | | | |
|---|---|---|---|---|---|---|---|
| Parameters | E2-I | E2-Q | E5B-I | E5B-Q | E6-I | E6-Q | GPS L1-CA |
| Native notation | B1 | B1 | B2 | B2 | B3 | B3 | --- |

| Code modulation | BPSK(2) | BPSK(2) | BPSK(2) | BPSK(10) | BPSK(10) | BPSK (10) | BPSK (1) |
|---|---|---|---|---|---|---|---|
| Carrier frequency, MHz | 1561.098 | 1561.098 | 1207.14 | 1207.14 | 1268.52 | 1268.52 | 1575.42 |
| Chip rate, Mchips/s | 2.046 | 2.046 | 2.046 | 10.230 | 10.230 | 10.230 | 1.023 |
| Code period, chips | 2046 | ?? | 2046 | ?? | 10230 | ?? | 1023 |
| Code period, ms | 1.0 | >400 | 1.0 | >160 | 1.0 | >160 | 1.0 |
| Symbols/s | 50 | ?? | 50 | ?? | 50 | ?? | 50 |
| Navigation frames, s | 6 | ?? | 6 | ?? | ?? | ?? | 6 |
| Navigation sub-frames, s | 30 | ?? | 30 | ?? | ?? | ?? | 30 |
| Navigation period, min | 12.0 | ?? | 12.0 | ?? | ?? | ?? | 12.5 |

Characteristics of the "I" signals on E2 and E5B are generally similar to the civilian codes of GPS (L1-CA and L2C), but Compass signals have somewhat greater power. The notation of Compass signals used in this page follows the naming of the frequency bands and agrees with the notation used in the American literature on the subject, but the notation used by the Chinese seems to be different and is quoted in the first row of the table.

## Operation

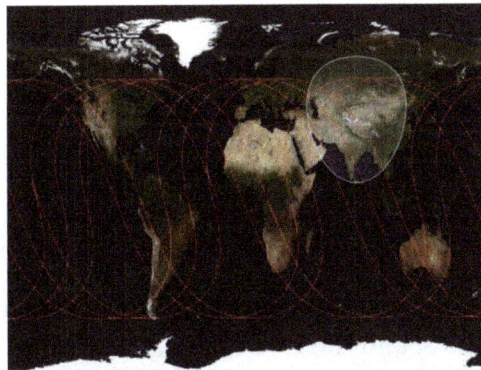

Ground track of BeiDou-M5 (2012-050A)

In December 2011, the system went into operation on a trial basis. It has started providing navigation, positioning and timing data to China and the neighbouring area for free from 27 December. During this trial run, Compass will offer positioning accuracy to within 25 meters, but the precision will improve as more satellites are launched. Upon the system's official launch, it pledged to offer general users positioning information accurate to the nearest 10 m, measure speeds within 0.2 m per second, and provide signals for clock synchronisation accurate to 0.02 microseconds.

The BeiDou-2 system began offering services for the Asia-Pacific region in December 2012. At this time, the system could provide positioning data between longitude 55°E to 180°E and from latitude 55°S to 55°N.

## Completion

In December 2011, Xinhua stated that "[t]he basic structure of the Beidou system has now been established, and engineers are now conducting comprehensive system test and evaluation. The system will provide test-run services of positioning, navigation and time for China and the neighboring areas before the end of this year, according to the authorities." The system became operational in the China region that same month. The global navigation system should be finished by 2020. As of December 2012, 16 satellites for BeiDou-2 have been launched, 14 of them are in service.

## Constellation

| Summary of satellites | | | | | |
|---|---|---|---|---|---|
| **Block** | **Launch Period** | **Satellite launches** | | | **Currently in orbit and healthy** |
| | | **Success** | **Failure** | **Planned** | |
| 1 | 2000–2007 | 4 | 0 | 0 | 0 |
| 2 | 2007-2012 | 16 | 0 | 0 | 14 |
| 3 | From 2015 | 6 | 0 | 18 | 6 |
| Total | | 26 | 0 | 18 | 20 |
| (Last update: March 29, 2016) | | | | | |

The regional Beidou-1 system was decommissioned at the end of 2012.

The first satellite of the second-generation system, Compass-M1 was launched in 2007. It was followed by further nine satellites during 2009-2011, achieving functional regional coverage. A total of 16 satellites were launched during this phase.

In 2015, the system began its transition towards global coverage with the first launch of a new-generation of satellites, and the 17th one within the new system.

On July 25, 2015, the 18th and 19th satellites were successfully launched from the Xichang Satellite Launch Center, marking the first time for China to launch two satellites at once on top of a Long March 3B/Expedition-1 carrier rocket. The Expedition-1 is an independent upper stage capable of delivering one or more spacecraft into different orbits.

The three latest satellites will jointly undergo testing of a new system of navigation signaling and inter-satellite links, and start providing navigation services when ready.

# Galileo (Satellite Navigation)

Galileo is the global navigation satellite system (GNSS) that is currently being created by the European Union (EU) through the European Space Agency (ESA) and the European GNSS Agency (GSA), headquartered in Prague in the Czech Republic, with two ground operations centres, Oberpfaffenhofen near Munich in Germany and Fucino in Italy. The €5 billion project is named after the Italian astronomer Galileo Galilei. One of the aims of Galileo is to provide an indigenous alternative high-precision positioning system upon which European nations can rely, independently from the Russian GLONASS, China Bei Dou and US GPS systems, in case they were disabled by their operators. The use of basic (low-precision) Galileo services will be free and open to everyone. The high-precision capabilities will be available for paying commercial users. Galileo is intended to provide horizontal and vertical position measurements within 1-metre precision, and better positioning services at high latitudes than other positioning systems.

Galileo is to provide a new global search and rescue (SAR) function as part of the MEOSAR system. Satellites will be equipped with a transponder which will relay distress signals from emergency beacons to the Rescue coordination centre, which will then initiate a rescue operation. At the same time, the system is projected to provide a signal, the Return Link Message (RLM), *to* the emergency beacon, informing them that their situation has been detected and help is on the way. This latter feature is new and is considered a major upgrade compared to the existing Cospas-Sarsat system, which do not provide feedback to the user. Tests in February 2014 found that for Galileo's search and rescue function, operating as part of the existing International Cospas-Sarsat Programme, 77% of simulated distress locations can be pinpointed within 2 km, and 95% within 5 km.

The first Galileo test satellite, the GIOVE-A, was launched 28 December 2005, while the first satellite to be part of the operational system was launched on 21 October 2011. As of May 2016 the system has 14 of 30 satellites in orbit. Galileo will start offering Early Operational Capability (EOC) from 2016, go to Initial Operational Capability (IOC) in 2017–18 and reach Full Operational Capability (FOC) in 2019. The complete 30-satellite Galileo system (24 operational and 6 active spares) is expected by 2020.

## History

Headquarters of the Galileo system in Prague

## Main Objectives

In 1999, the different concepts of the three main contributors of ESA (Germany, France and Italy) for Galileo were compared and reduced to one by a joint team of engineers from all three countries. The first stage of the Galileo programme was agreed upon officially on 26 May 2003 by the European Union and the European Space Agency. The system is intended primarily for civilian use, unlike the more military-orientated systems of the United States (GPS), Russia (GLONASS), and China (Beidou-1/2, COMPASS). The European system will only be subject to shutdown for military purposes in extreme circumstances (like armed conflict). It will be available at its full precision to both civil and military users.

## Funding

The European Commission had some difficulty funding the project's next stage, after several allegedly "per annum" sales projection graphs for the project were exposed in November 2001 as "cumulative" projections which for each year projected included all previous years of sales. The attention that was brought to this multibillion-euro growing error in sales forecasts resulted in a general awareness in the Commission and elsewhere that it was unlikely that the program would yield the return on investment that had previously been suggested to investors and decision-makers. On 17 January 2002, a spokesman for the project stated that, as a result of US pressure and economic difficulties, "Galileo is almost dead."

A few months later, however, the situation changed dramatically. European Union member states decided it was important to have a satellite-based positioning and timing infrastructure that the US could not easily turn off in times of political conflict.

The European Union and the European Space Agency agreed in March 2002 to fund the project, pending a review in 2003 (which was completed on 26 May 2003). The starting cost for the period ending in 2005 is estimated at €1.1 billion. The required satellites (the planned number is 30) were to be launched between 2011 and 2014, with the system up and running and under civilian control from 2019. The final cost is estimated at €3 billion, including the infrastructure on Earth, constructed in 2006 and 2007. The plan was for private companies and investors to invest at least two-thirds of the cost of implementation, with the EU and ESA dividing the remaining cost. The base *Open Service* is to be available without charge to anyone with a Galileo-compatible receiver, with an encrypted higher-bandwidth improved-precision *Commercial Service* available at a cost. By early 2011 costs for the project had run 50% over initial estimates.

The German Aerospace Center (DLR) contributes the largest portion of the Galileo funds, and is crucial in the development and application of the system with its facilities of the Earth Observation Center, and the Institute for Communication and Navigation in Neustrelitz.

## Tension with the United States

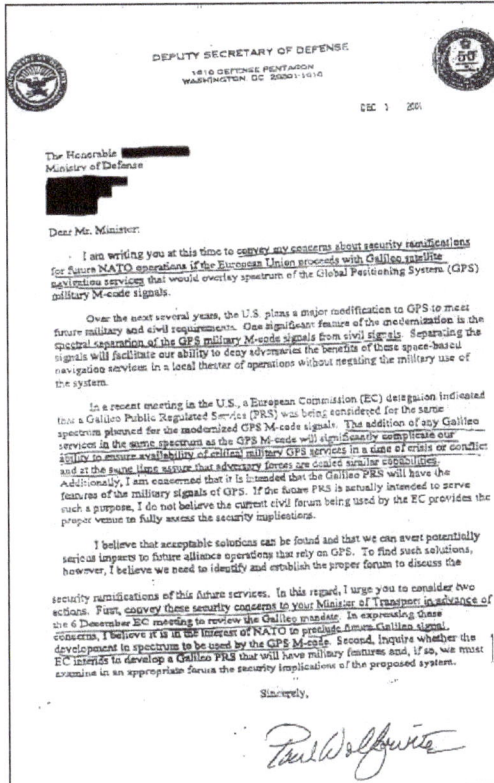

A December 2001 letter from U.S. Deputy Secretary of Defense Paul Wolfowitz to the Ministers of the EU states as part of the US-lobbying campaign against Galileo

Galileo is intended to be an EU civilian GNSS that allows all users access to it. GPS is a US military GNSS that provides location signals that have high precision to US military users, while also providing less precise location signals to others. The GPS had the capability to block the "civilian" signals while still being able to use the "military" signal (M-band). A primary motivation for the Galileo project was the EU concern that the US could deny others access to GPS during political disagreements.

Since Galileo was designed to provide the highest possible precision (greater than GPS) to anyone, the US was concerned that an enemy could use Galileo signals in military strikes against the US and its allies (some weapons like missiles use GNSSs for guidance). The frequency initially chosen for Galileo would have made it impossible for the US to block the Galileo signals without also interfering with its own GPS signals. The US did not want to lose their GNSS capability with GPS while denying enemies the use of GNSS. Some US officials became especially concerned when Chinese interest in Galileo was reported.

An anonymous EU official claimed that the US officials implied that they might consider shooting down Galileo satellites in the event of a major conflict in which Galileo was used in attacks against American forces. The EU's stance is that Galileo is a neutral technology, available to all countries and everyone. At first, EU officials did not want to change their original plans for Galileo, but have since reached a compromise, that Galileo was to use a different frequency. This allowed the blocking or jamming of either GNSS without affecting the other (jam Galileo without affecting GPS, or jam GPS but not Galileo), giving the US a greater advantage in conflicts in which it has the electronic warfare upper hand.

## GPS and Galileo

One of the reasons given for developing Galileo as an independent system was that position information from GPS can be made significantly inaccurate by the deliberate application of universal Selective Availability (SA) by the US military. GPS is widely used worldwide for civilian applications; Galileo's proponents argued that civil infrastructure, including aeroplane navigation and landing, should not rely solely upon a system with this vulnerability.

On 2 May 2000, SA was disabled by the President of the United States, Bill Clinton; in late 2001 the entity managing the GPS confirmed that they did not intend to enable selective availability ever again. Though Selective Availability capability still exists, on 19 September 2007 the US Department of Defense announced that newer GPS satellites would not be capable of implementing Selective Availability; the wave of Block IIF satellites launched in 2009, and all subsequent GPS satellites, are stated not to support SA. As old satellites are replaced in the GPS Block IIIA program, SA will cease to be an option. The modernisation programme also contains standardised features that allow GPS III and Galileo systems to inter-operate, allowing receivers to be developed to utilise GPS and Galileo together to create an even more precise GNSS.

## Cooperation with the United States

In June 2004, in a signed agreement with the United States, the European Union agreed to switch to a modulation known as BOC(1,1) (Binary Offset Carrier 1.1) allowing the coexistence of both GPS and Galileo, and the future combined use of both systems.

The European Union also agreed to address the "mutual concerns related to the protection of allied and US national security capabilities."

## First Experimental Satellites: GIOVE-A and GIOVE-B

The first experimental satellite, GIOVE-A, was launched in December 2005 and was followed by a second test satellite, GIOVE-B, launched in April 2008. After successful completion of the In-Orbit Validation (IOV) phase, additional satellites were launched. On 30 November 2007 the 27 EU transportation ministers involved reached an agreement that Galileo should be operational by 2013, but later press releases suggest it was delayed to 2014.

## Funding Again, Governance Issues

In mid-2006 the public/private partnership fell apart, and the European Commission decided to nationalise the Galileo programme.

In early 2007 the EU had yet to decide how to pay for the system and the project was said to be "in deep crisis" due to lack of more public funds. German Transport Minister Wolfgang Tiefensee was particularly doubtful about the consortium's ability to end the infighting at a time when only one testbed satellite had been successfully launched.

Although a decision was yet to be reached, on 13 July 2007 EU countries discussed cutting €548m ($755m, £370m) from the union's competitiveness budget for the following year and shifting some of these funds to other parts of the financing pot, a move that could meet part of the cost of the union's Galileo satellite navigation system. European Union research and development projects could be scrapped to overcome a funding shortfall.

In November 2007, it was agreed to reallocate funds from the EU's agriculture and administration budgets and to soften the tendering process in order to invite more EU companies.

In April 2008, the EU transport ministers approved the Galileo Implementation Regulation. This allowed the €3.4bn to be released from the EU's agriculture and administration budgets to allow the issuing of contracts to start construction of the ground station and the satellites.

In June 2009, the European Court of Auditors published a report, pointing out governance issues, substantial delays and budget overruns that led to project stalling in 2007, leading to further delays and failures.

In October 2009, the European Commission cut the number of satellites definitively planned from 28 to 22, with plans to order the remaining six at a later time. It also announced that the first OS, PRS and SoL signal would be available in 2013, and the CS and SOL some time later. The €3.4 billion budget for the 2006–2013 period was considered insufficient. In 2010 the think-tank Open Europe estimated the total cost of Galileo from start to 20 years after completion at €22.2 billion, borne entirely by taxpayers. Under the original estimates made in 2000, this cost would have been €7.7 billion, with €2.6 billion borne by taxpayers and the rest by private investors.

In November 2009, a ground station for Galileo was inaugurated near Kourou (French Guiana).

The launch of the first four in-orbit validation (IOV) satellites was planned for the second half of 2011, and the launch of full operational capability (FOC) satellites was planned to start in late 2012.

In March 2010 it was verified that the budget for Galileo would only be available to provide the 4 IOV and 14 FOC satellites by 2014, with no funds then committed to bring the constellation above this 60% capacity. Paul Verhoef, the satellite navigation program manager at the European Commission, indicated that this limited funding would have serious consequences commenting at one point "To give you an idea, that would mean that for three weeks in the year you will not have satellite navigation" in reference to the proposed 18-vehicle constellation.

In July 2010, the European Commission estimated further delays and additional costs of the project to grow up to €1.5-€1.7 billion, and moved the estimated date of completion to 2018. After completion the system will need to be subsidised by governments at €750 million per year. An additional €1.9 billion was planned to be spent bringing the system up to the full complement of 30 satellites (27 operational + 3 active spares).

In December 2010, EU ministers in Brussels voted Prague, in the Czech Republic, as the headquarters of the Galileo project.

In January 2011, infrastructure costs up to 2020 were estimated at €5.3 billion. In that same month, Wikileaks revealed that Berry Smutny, the CEO of the German satellite company OHB-System, said that Galileo "is a stupid idea that primarily serves French interests". The BBC understood in 2011 that €500 million (£440M) would become available to make the extra purchase, taking Galileo within a few years from 18 operational satellites to 24.

Galileo launch on a Soyuz rocket, 21 October 2011

The first two Galileo In-Orbit Validation satellites were launched by Soyuz ST-B flown from Guiana Space Centre on 21 October 2011, and the remaining two on 12 October 2012.

22 further satellites with Full Operational Capability (FOC) were on order as of 2012. The first 4 pairs of satellites were launched on 22 August 2014, 27 March 2015, 11 September 2015 and 17 December 2015.

## International Involvement

In September 2003, China joined the Galileo project. China was to invest €230 million (US$302 million, GBP 155 million, CNY 2.34 billion) in the project over the following years.

In July 2004, Israel signed an agreement with the EU to become a partner in the Galileo project.

On 3 June 2005 the EU and Ukraine signed an agreement for Ukraine to join the project, as noted in a press release.

As of November 2005, Morocco also joined the programme.

In Mid-2006, the Public-Private Partnership fell apart and the European Commission decided to nationalise Galileo as an EU programme.

In November 2006, China opted instead to independently develop the Beidou navigation system satellite navigation system.

On 30 November 2007, the 27 member states of the European Union unanimously agreed to move forward with the project, with plans for bases in Germany and Italy. Spain did not approve during the initial vote, but approved it later that day. This greatly improves the viability of the Galileo project: "The EU's executive had previously said that if agreement was not reached by January 2008, the long-troubled project would essentially be dead."

On 3 April 2009, Norway too joined the programme pledging €68.9 million toward development costs and allowing its companies to bid for the construction contracts. Norway, while not a member of the EU, is a member of ESA.

On 18 December 2013, Switzerland signed a cooperation agreement to fully participate in the program, and retroactively contributed €80 million for the period 2008-2013. As a member of ESA, it already collaborated in the development of the Galileo satellites, contributing the state-of-the-art hydrogen-maser clocks. Switzerland's financial commitment for the period 2014-2020 will be calculated in accordance with the standard formula applied for the Swiss participation in the EU research Framework Programme.

## System Description

### Space Segment

As of 2012, the system is scheduled to reach full operation in 2020 with the following specifications:

- 30 in-orbit spacecraft (24 in full service and 6 spares)

- Orbital altitude: 23,222 km (MEO)

- 3 orbital planes, 56° inclination, ascending nodes separated by 120° longitude (8 operational satellites and 2 active spares per orbital plane)

- Satellite lifetime: >12 years

- Satellite mass: 675 kg

- Satellite body dimensions: 2.7 m × 1.2 m × 1.1 m

- Span of solar arrays: 18.7 m

- Power of solar arrays: 1.5 kW (end of life)

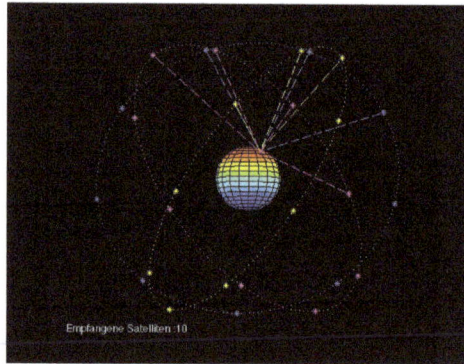

Constellation visibility from a location on Earth's surface (animation)

## Ground Segment

The system's orbit and signal accuracy is controlled by a ground segment consisting of:

- 1 ground control centre, located in Oberpfaffenhofen

- 1 ground mission centre, located in Fucino

- 5 tracking stations, located in Kiruna, Kourou, Noumea, Sainte-Marie, Réunion & Redu

- Several uplink stations

- Several sensor stations

- A data dissemination network between stations

## Services

The Galileo system will have five main services:

Open access navigation

> This will be available without charge for use by anyone with appropriate mass-market equipment; simple timing, and positioning down to 1 metre.

Commercial navigation (encrypted)

> High precision to the centimetre; guaranteed service for which service providers will charge fees.

Safety of life navigation

> Open service; for applications where guaranteed precision is essential. Integrity messages will warn of errors.

Public regulated navigation (encrypted)

> Continuous availability even if other services are disabled in time of crisis; Government agencies will be main users.

Search and rescue

> System will pick up distress beacon locations; feasible to send feedback, e.g. confirming help is on its way.

Other secondary services will also be available.

## Concept

Each satellite will have two rubidium atomic clocks and two passive hydrogen maser atomic clocks, critical to any satellite-navigation system, and a number of other components. The clocks will provide an accurate timing signal to allow a receiver to calculate the time that it takes the signal to reach it. This information is used to calculate the position of the receiver by trilaterating the difference in received signals from multiple satellites.

## Constellation

| Summary of satellites | | | | |
|---|---|---|---|---|
| **Block** | **Launch Period** | **Satellite launches** | | | **Currently in operational orbit and healthy** |
| | | **Full success** | **Failure** | **Planned** | |
| GIOVE | 2005–2008 | 2 | 0 | 0 | 0 |
| IOV | 2011–2012 | 4 | 0 | 0 | 3 |

| FOC | From 2014 | 8 | 2* | 18 | 8 |
|-----|-----------|---|----|----|----|
| Total | | 16 | 0 | 18 | 11 |
| * Two partial launch failures resulting in 2 satellites orbiting in a degraded orbit | | | | | |

## Galileo Satellite Test Beds: GIOVE

GIOVE-A was successfully launched 28 December 2005

In 2004 the Galileo System Test Bed Version 1 (GSTB-V1) project validated the on-ground algorithms for Orbit Determination and Time Synchronisation (OD&TS). This project, led by ESA and European Satellite Navigation Industries, has provided industry with fundamental knowledge to develop the mission segment of the Galileo positioning system.

- GIOVE-A is the first GIOVE (Galileo In-Orbit Validation Element) test satellite. It was built by Surrey Satellite Technology Ltd (SSTL), and successfully launched on 28 December 2005 by the European Space Agency and the Galileo Joint. Operation of GIOVE-A ensured that Galileo meets the frequency-filing allocation and reservation requirements for the International Telecommunication Union (ITU), a process that was required to be complete by June 2006.

- GIOVE-B, built by Astrium and Thales Alenia Space, has a more advanced payload than GIOVE-A. It was successfully launched on 27 April 2008 at 22:16 UTC (4.16 am Baikonur time) aboard a Soyuz-FG/Fregat rocket provided by Starsem.

A third satellite, GIOVE-A2, was originally planned to be built by SSTL for launch in the second half of 2008. Construction of GIOVE-A2 was terminated due to the successful launch and in-orbit operation of GIOVE-B.

The GIOVE Mission segment operated by European Satellite Navigation Industries used the GIOVE-A/B satellites to provide experimental results based on real data to be used for risk mitigation for the IOV satellites that followed on from the testbeds.

ESA organised the global network of ground stations to collect the measurements of GIOVE-A/B with the use of the GETR receivers for further systematic study. GETR receivers are supplied by Septentrio as well as the first Galileo navigation receivers to be used to test the functioning of the system at further stages of its deployment. Signal analysis of GIOVE-A/B data confirmed successful operation of all the Galileo signals with the tracking performance as expected.

## In-orbit Validation (IOV) Satellites

These testbed satellites were followed by four IOV Galileo satellites that are much closer to the final Galileo satellite design. The Search & Rescue feature is also installed. The first two satellites were launched on 21 October 2011 from Guiana Space Centre using a Soyuz launcher, the other two on 12 October 2012. This enables key validation tests, since earth-based receivers such as those in cars and phones need to "see" a minimum of four satellites in order to calculate their position in three dimensions. Those 4 IOV Galileo satellites were constructed by Astrium GmbH and Thales Alenia Space. On 12 March 2013, a first fix was performed using those four IOV satellites. Once this In-Orbit Validation (IOV) phase has been completed, the remaining satellites will be installed to reach the Full Operational Capability.

## Full Operational Capability (FOC) Satellites

On 7 January 2010, it was announced that the contract to build the first 14 FOC satellites was awarded to OHB System and Surrey Satellite Technology Limited (SSTL). Fourteen satellites will be built at a cost of €566M (£510M; $811M). Arianespace will launch the satellites for a cost of €397M (£358M; $569M). The European Commission also announced that the €85 million contract for system support covering industrial services required by ESA for integration and validation of the Galileo system had been awarded to Thales Alenia Space. Thales Alenia Space subcontract performances to Astrium GmbH and security to Thales Communications.

In February 2012, an additional order of eight satellites was awarded to OHB Systems for €250M ($327M), after outbidding EADS Astrium tender offer. Thus bringing the total to 22 FOC satellites.

On 7 May 2014, the first two FOC satellites landed in Guyana for their joint launch planned in summer Originally planned for launch during 2013, problems tooling and establishing the production line for assembly led to a delay of a year in serial production of Galileo satellites. These two satellites (Galileo satellites GSAT0201 and GSAT0202) were launched on 22 August 2014. The names of these satellites are Doresa and Milena named after European children who had previously won a drawing contest. On 23 August 2014, launch service provider Arianespace announced that the flight VS09 experienced anomaly and satellites were injected into an incorrect orbit.

Satellites GSAT0203 and GSAT0204 were launched successfully on 27 March 2015 from Guiana Space Centre using a Soyuz four stage launcher. Using the same Soyuz launcher and launchpad, satellites GSAT0205 and GSAT0206 were launched successfully on 11 September 2015.

Satellites GSAT0208 and GSAT0209 were successfully launched from Kourou, French Guiana, using the Soyuz launcher on December 17, 2015.

Starting in 2016, deployment of the last twelve satellites will use a modified Ariane 5 launcher, named Ariane 5 ES, capable of placing four Galileo satellites into orbit per launch.

Satellites GSAT0210 and GSAT0211 were launched on 24 May 2016 and are being commissioned. The next four are planned for launch in November 2016 on an Ariane 5 ES.

## Applications and Impact

## Science Projects Using Galileo

In July 2006 an international consortium of universities and research institutions embarked on a study of potential scientific applications of the Galileo constellation. This project, named GEO6, is a broad study oriented to the general scientific community, aiming to define and implement new applications of Galileo.

Among the various GNSS users identified by the Galileo Joint Undertaking, the GEO6, project addresses the Scientific User Community (UC).

The GEO6 project aims at fostering possible novel applications within the scientific UC of GNSS signals, and particularly of Galileo.

The AGILE project is an EU-funded project devoted to the study of the technical and commercial aspects of location-based services (LBS). It includes technical analysis of the benefits brought by Galileo (and EGNOS) and studies the hybridisation of Galileo with other positioning technologies (network-based, WLAN, etc.). Within these project, some pilot prototypes were implemented and demonstrated.

On the basis of the potential number of users, potential revenues for Galileo Operating Company or Concessionaire (GOC), international relevance, and level of innovation, a set of Priority Applications (PA) will be selected by the consortium and developed within the time-frame of the same project.

These applications will help to increase and optimise the use of the EGNOS services and the opportunities offered by the Galileo Signal Test-Bed (GSTB-V2) and the Galileo (IOV) phase.

## Coins

The European Satellite Navigation project was selected as the main motif of a very high value collectors' coin: the Austrian European Satellite Navigation commemorative coin, minted on 1 March 2006. The coin has a silver ring and gold-brown niobium "pill". In the reverse, the niobium portion depicts navigation satellites orbiting the Earth. The ring shows different modes of transport, an aeroplane, a car, a container ship, a train and a lorry, for which satellite navigation was developed.

Austrian €25 European Satellite Navigation commemorative coin, back

## References

- Johnson, Chalmers (2008). Nemesis: The Last Days of the American Republic. Holt. p. 235. ISBN 0-8050-8728-1.

- "China launches Long March 3B rocket with Beidou-3 navigation satellite". Spaceflight Insider. Retrieved 2 March 2016.

- "China successfully launched the first New-Generation Beidou Navigation Satellite". BeiDou. 1 April 2015.

- "On a Civil Global Navigation Satellite System (GNSS) between the European Community and its Member States and Ukraine" (PDF). Retrieved 12 January 2015.

- "Arianespace continues deployment of Galileo, a flagship project for Europe" (PDF). Arianespace. March 2015. Retrieved 2015-05-31.

- "Galileo pair preparing for December launch". European Space Agency. 2 November 2015. Retrieved 13 December 2015.

- Correspondent, Jonathan Amos BBC Science. "Two more Galileo satellites launched". BBC News. Retrieved 2015-12-17.

- "Commission awards major contracts to make Galileo operational early 2014". 7 January 2010. Retrieved 19 April 2010.

- Rhian, Jason (22 August 2014). "Doresa and Milena Galileo spacecraft rise into morning sky via Soyuz ST-B". Spaceflight Insider.

- "Galileo satellites experience orbital injection anomaly on Soyuz launch: Initial report" (Press release). 23 August 2014. Archived from the original on 27 August 2014. Retrieved 27 August 2014.

- "Precise orbit determination of Beidou Satellites with precise positioning". Science China. 2012. Retrieved 26 June 2013.

- "US Department of Defense Reports on China's Space Capabilities". Space Safety Magazine. 27 May 2013. Retrieved 26 June 2013.

- Switzerland joins the EU's Galileo satellite navigation programme europa.eu Press release, 18 December 2013.

# 4

# Global Positioning System: A Comprehensive Study

Global positioning system (GPS) is a global satellite navigation system that provides the user with precise coordinates of locations present anywhere on the earth. This chapter studies the global positioning system thoroughly by exploring its components like GPS satellite blocks, GPS signals, GPS navigation device, GNSS positioning calculation, GNSS enhancement and error analysis for the global positioning system. All the main components of global positioning system are explained here.

## Global Positioning System

A U.S. Air Force Senior Airman runs through a checklist
during Global Positioning System satellite operations.

The Global Positioning System (GPS), also known as Navstar, is a global navigation satellite system (GNSS) that provides location and time information in all weather conditions, anywhere on or near the Earth where there is an unobstructed line of sight to four or more GPS satellites. The GPS system operates independently of any telephonic or internet reception, though these technologies can enhance the usefulness of the GPS positioning information. The GPS system provides critical positioning capabilities to military, civil, and commercial users around the world. The United States government created the system, maintains it, and makes it freely accessible to anyone with a GPS receiver.

The GPS project was launched in the United States in 1973 to overcome the limitations of previous navigation systems, integrating ideas from several predecessors, including a number of classified engineering design studies from the 1960s. The U.S. Department of Defense (DoD) developed the system, which originally used 24 satellites. It became fully operational in 1995. Roger L. Easton, Ivan A. Getting and Bradford Parkinson of the Applied Physics Laboratory are credited with inventing it.

Advances in technology and new demands on the existing system have now led to efforts to modernize the GPS and implement the next generation of GPS Block IIIA satellites and Next Generation Operational Control System (OCX). Announcements from Vice President Al Gore and the White House in 1998 initiated these changes. In 2000, the U.S. Congress authorized the modernization effort, GPS III.

In addition to GPS, other systems are in use or under development. The Russian Global Navigation Satellite System (GLONASS) was developed contemporaneously with GPS, but suffered from incomplete coverage of the globe until the mid-2000s. There are also the planned European Union Galileo positioning system, China's BeiDou Navigation Satellite System, the Japanese Quasi-Zenith Satellite System, and India's Indian Regional Navigation Satellite System.

## History

The design of GPS is based partly on similar ground-based radio-navigation systems, such as LORAN and the Decca Navigator, developed in the early 1940s and used by the British Royal Navy during World War II.

In 1956, the German-American physicist Friedwardt Winterberg proposed a test of general relativity — detecting time slowing in a strong gravitational field using accurate atomic clocks placed in orbit inside artificial satellites.

Special and general relativity predict that the clocks on the GPS satellites would be seen by the Earth's observers to run 38 microseconds faster per day than the clocks on the Earth. The GPS calculated positions would quickly drift into error, accumulating to 10 kilometers per day. The relativistic time effect of the GPS clocks running faster than the clocks on earth was corrected for in the design of GPS.

## Predecessors

The Soviet Union launched the first man-made satellite, Sputnik 1, in 1957. Two American physicists, William Guier and George Weiffenbach, at Johns Hopkins's Applied Physics Laboratory (APL), decided to monitor Sputnik's radio transmissions. Within hours they realized that, because of the Doppler effect, they could pinpoint where the satellite was along its orbit. The Director of the APL gave them access to their UNIVAC to do the heavy calculations required.

The next spring, Frank McClure, the deputy director of the APL, asked Guier and Weiffenbach to investigate the inverse problem — pinpointing the user's location, given that of the satellite. (At the time, the Navy was developing the submarine-launched Polaris missile, which required them to know the submarine's location.) This led them and APL to develop the TRANSIT system. In 1959, ARPA (renamed DARPA in 1972) also played a role in TRANSIT.

Official logo for NAVSTAR GPS

Emblem of the 50th Space Wing

The first satellite navigation system, TRANSIT, used by the United States Navy, was first successfully tested in 1960. It used a constellation of five satellites and could provide a navigational fix approximately once per hour.

In 1967, the U.S. Navy developed the Timation satellite that proved the ability to place accurate clocks in space, a technology required by GPS.

In the 1970s, the ground-based OMEGA navigation system, based on phase comparison of signal transmission from pairs of stations, became the first worldwide radio navigation system. Limitations of these systems drove the need for a more universal navigation solution with greater accuracy.

While there were wide needs for accurate navigation in military and civilian sectors, almost none of those was seen as justification for the billions of dollars it would cost in research, development, deployment, and operation for a constellation of navigation satellites. During the Cold War arms race, the nuclear threat to the existence of the United States was the one need that did justify this cost in the view of the United States Congress. This deterrent effect is why GPS was funded. It is also the reason for the ultra secrecy at that time. The nuclear triad consisted of the United States Navy's submarine-launched ballistic missiles (SLBMs) along with United States Air Force (USAF) strategic bombers and intercontinental ballistic missiles (ICBMs). Considered vital to the nuclear deterrence posture, accurate determination of the SLBM launch position was a force multiplier.

Precise navigation would enable United States ballistic missile submarines to get an accurate fix of their positions before they launched their SLBMs. The USAF, with two thirds of the nuclear triad, also had requirements for a more accurate and reliable navigation system. The Navy and Air Force were developing their own technologies in parallel to solve what was essentially the same problem.

To increase the survivability of ICBMs, there was a proposal to use mobile launch platforms (such as Russian SS-24 and SS-25) and so the need to fix the launch position had similarity to the SLBM situation.

In 1960, the Air Force proposed a radio-navigation system called MOSAIC (MObile System for Accurate ICBM Control) that was essentially a 3-D LORAN. A follow-on study, Project 57, was worked in 1963 and it was "in this study that the GPS concept was born." That same year, the concept was pursued as Project 621B, which had "many of the attributes that you now see in GPS" and promised increased accuracy for Air Force bombers as well as ICBMs.

Updates from the Navy TRANSIT system were too slow for the high speeds of Air Force operation. The Naval Research Laboratory continued advancements with their Timation (Time Navigation) satellites, first launched in 1967, and with the third one in 1974 carrying the first atomic clock into orbit.

Another important predecessor to GPS came from a different branch of the United States military. In 1964, the United States Army orbited its first Sequential Collation of Range (SECOR) satellite used for geodetic surveying. The SECOR system included three ground-based transmitters from known locations that would send signals to the satellite transponder in orbit. A fourth ground-based station, at an undetermined position, could then use those signals to fix its location precisely. The last SECOR satellite was launched in 1969.

Decades later, during the early years of GPS, civilian surveying became one of the first fields to make use of the new technology, because surveyors could reap benefits of signals from the less-than-complete GPS constellation years before it was declared operational. GPS can be thought of as an evolution of the SECOR system where the ground-based transmitters have been migrated into orbit.

## Development

With these parallel developments in the 1960s, it was realized that a superior system could be developed by synthesizing the best technologies from 621B, Transit, Timation, and SECOR in a multi-service program.

During Labor Day weekend in 1973, a meeting of about twelve military officers at the Pentagon discussed the creation of a *Defense Navigation Satellite System (DNSS)*. It was at this meeting that the real synthesis that became GPS was created. Later that year, the DNSS program was named *Navstar*, or Navigation System Using Timing and Ranging. With the individual satellites being associated with the name Navstar (as with the predecessors Transit and Timation), a more fully encompassing name was used to identify the constellation of Navstar satellites, *Navstar-GPS*. Ten "Block I" prototype satellites were launched between 1978 and 1985 (an additional unit was destroyed in a launch failure).

After Korean Air Lines Flight 007, a Boeing 747 carrying 269 people, was shot down in 1983 after straying into the USSR's prohibited airspace, in the vicinity of Sakhalin and Moneron Islands, President Ronald Reagan issued a directive making GPS freely available for civilian use, once it was sufficiently developed, as a common good. The first Block II satellite was launched on February 14, 1989, and the 24th satellite was launched in 1994. The GPS program cost at this point, not including the cost of the user equipment, but including the costs of the satellite launches, has been estimated at about USD$5 billion (then-year dollars). Roger L. Easton is widely credited as the primary inventor of GPS.

Initially, the highest quality signal was reserved for military use, and the signal available for civilian use was intentionally degraded (Selective Availability). This changed with President Bill Clinton signing a policy directive in 1996 to turn off Selective Availability in May 2000 to provide the same precision to civilians that was afforded to the military. The directive was proposed by the U.S. Secretary of Defense, William Perry, because of the widespread growth of differential GPS services to improve civilian accuracy and eliminate the U.S. military advantage. Moreover, the U.S. military was actively developing technologies to deny GPS service to potential adversaries on a regional basis.

Since its deployment, the U.S. has implemented several improvements to the GPS service including new signals for civil use and increased accuracy and integrity for all users, all the while maintaining compatibility with existing GPS equipment. Modernization of the satellite system has been an ongoing initiative by the U.S. Department of Defense through a series of satellite acquisitions to meet the growing needs of the military, civilians, and the commercial market.

As of early 2015, high-quality, FAA grade, Standard Positioning Service (SPS) GPS receivers provide horizontal accuracy of better than 3.5 meters, although many factors such as receiver quality and atmospheric issues can affect this accuracy.

GPS is owned and operated by the United States Government as a national resource. The Department of Defense is the steward of GPS. *Interagency GPS Executive Board (IGEB)* oversaw GPS policy matters from 1996 to 2004. After that the National Space-Based Positioning, Navigation and Timing Executive Committee was established by presidential directive in 2004 to advise and coordinate federal departments and agencies on matters concerning the GPS and related systems. The executive committee is chaired jointly by the deputy secretaries of defense and transportation. Its membership includes equivalent-level officials from the departments of state, commerce, and homeland security, the joint chiefs of staff, and NASA. Components of the executive office of the president participate as observers to the executive committee, and the FCC chairman participates as a liaison.

The U.S. Department of Defense is required by law to "maintain a Standard Positioning Service (as defined in the federal radio navigation plan and the standard positioning service signal specification) that will be available on a continuous, worldwide basis," and "develop measures to prevent hostile use of GPS and its augmentations without unduly disrupting or degrading civilian uses."

## Timeline and Modernization

| Summary of satellites | | | | | | |
|---|---|---|---|---|---|---|
| **Block** | **Launch period** | **Satellite launches** | | | | **Currently in orbit and healthy** |
| | | **Success** | **Failure** | **In preparation** | **Planned** | |
| I | 1978–1985 | 10 | 1 | 0 | 0 | 0 |
| II | 1989–1990 | 9 | 0 | 0 | 0 | 0 |
| IIA | 1990–1997 | 19 | 0 | 0 | 0 | 0 |
| IIR | 1997–2004 | 12 | 1 | 0 | 0 | 12 |
| IIR-M | 2005–2009 | 8 | 0 | 0 | 0 | 7 |
| IIF | 2010–2016 | 12 | 0 | 0 | 0 | 12 |
| IIIA | From 2017 | 0 | 0 | 0 | 12 | 0 |
| IIIB | — | 0 | 0 | 0 | 8 | 0 |
| IIIC | — | 0 | 0 | 0 | 16 | 0 |
| Total | | 70 | 2 | 0 | 36 | 31 |

(Last update: March 9, 2016)
8 satellites from Block IIA are placed in reserve
USA-203 from Block IIR-M is unhealthy

- In 1972, the USAF Central Inertial Guidance Test Facility (Holloman AFB), conducted developmental flight tests of two prototype GPS receivers over White Sands Missile Range, using ground-based pseudo-satellites.

- In 1978, the first experimental Block-I GPS satellite was launched.

- In 1983, after Soviet interceptor aircraft shot down the civilian airliner KAL 007 that strayed into prohibited airspace because of navigational errors, killing all 269 people on board, U.S. President Ronald Reagan announced that GPS would be made available for civilian uses once it was completed, although it had been previously published [in Navigation magazine] that the CA code (Coarse Acquisition code) would be available to civilian users.

- By 1985, ten more experimental Block-I satellites had been launched to validate the concept.

- Beginning in 1988, Command & Control of these satellites was transitioned from Onizuka AFS, California to the 2nd Satellite Control Squadron (2SCS) located at Falcon Air Force Station in Colorado Springs, Colorado.

- On February 14, 1989, the first modern Block-II satellite was launched.

- The Gulf War from 1990 to 1991 was the first conflict in which the military widely used GPS.

- In 1991, a project to create a miniature GPS receiver successfully ended, replacing the previous 23 kg military receivers with a 1.25 kg handheld receiver.

- In 1992, the 2nd Space Wing, which originally managed the system, was inactivated and replaced by the 50th Space Wing.

- By December 1993, GPS achieved initial operational capability (IOC), indicating a full constellation (24 satellites) was available and providing the Standard Positioning Service (SPS).

- Full Operational Capability (FOC) was declared by Air Force Space Command (AFSPC) in April 1995, signifying full availability of the military's secure Precise Positioning Service (PPS).

- In 1996, recognizing the importance of GPS to civilian users as well as military users, U.S. President Bill Clinton issued a policy directive declaring GPS a dual-use system and establishing an Interagency GPS Executive Board to manage it as a national asset.

- In 1998, United States Vice President Al Gore announced plans to upgrade GPS with two new civilian signals for enhanced user accuracy and reliability, particularly with respect to aviation safety and in 2000 the United States Congress authorized the effort, referring to it as *GPS III*.

- On May 2, 2000 "Selective Availability" was discontinued as a result of the 1996 executive order, allowing users to receive a non-degraded signal globally.

- In 2004, the United States Government signed an agreement with the European Community establishing cooperation related to GPS and Europe's planned Galileo system.

- In 2004, United States President George W. Bush updated the national policy and replaced the executive board with the National Executive Committee for Space-Based Positioning, Navigation, and Timing.

- November 2004, Qualcomm announced successful tests of assisted GPS for mobile phones.

- In 2005, the first modernized GPS satellite was launched and began transmitting a second civilian signal (L2C) for enhanced user performance.

- On September 14, 2007, the aging mainframe-based Ground Segment Control System was transferred to the new Architecture Evolution Plan.

- On May 19, 2009, the United States Government Accountability Office issued a report warning that some GPS satellites could fail as soon as 2010.

- On May 21, 2009, the Air Force Space Command allayed fears of GPS failure saying "There's only a small risk we will not continue to exceed our performance standard."

- On January 11, 2010, an update of ground control systems caused a software incompatibility with 8000 to 10000 military receivers manufactured by a division of Trimble Navigation Limited of Sunnyvale, Calif.

- On February 25, 2010, the U.S. Air Force awarded the contract to develop the GPS Next Generation Operational Control System (OCX) to improve accuracy and availability of GPS navigation signals, and serve as a critical part of GPS modernization.

## Awards

On February 10, 1993, the National Aeronautic Association selected the GPS Team as winners of the 1992 Robert J. Collier Trophy, the nation's most prestigious aviation award. This team combines researchers from the Naval Research Laboratory, the USAF, the Aerospace Corporation, Rockwell International Corporation, and IBM Federal Systems Company. The citation honors them "for the most significant development for safe and efficient navigation and surveillance of air and spacecraft since the introduction of radio navigation 50 years ago."

Two GPS developers received the National Academy of Engineering Charles Stark Draper Prize for 2003:

- Ivan Getting, emeritus president of The Aerospace Corporation and an engineer at the Massachusetts Institute of Technology, established the basis for GPS, improving on the World War II land-based radio system called LORAN (*Long-range Radio Aid to Navigation*).

- Bradford Parkinson, professor of aeronautics and astronautics at Stanford University, conceived the present satellite-based system in the early 1960s and developed it in conjunction with the U.S. Air Force. Parkinson served twenty-one years in the Air Force, from 1957 to 1978, and retired with the rank of colonel.

GPS developer Roger L. Easton received the National Medal of Technology on February 13, 2006.

Francis X. Kane (Col. USAF, ret.) was inducted into the U.S. Air Force Space and Missile Pioneers Hall of Fame at Lackland A.F.B., San Antonio, Texas, March 2, 2010 for his role in space technology development and the engineering design concept of GPS conducted as part of Project 621B.

In 1998, GPS technology was inducted into the Space Foundation Space Technology Hall of Fame.

On October 4, 2011, the International Astronautical Federation (IAF) awarded the Global Positioning System (GPS) its 60th Anniversary Award, nominated by IAF member, the American Institute for Aeronautics and Astronautics (AIAA). The IAF Honors and Awards Committee recognized the uniqueness of the GPS program and the exemplary role it has played in building international collaboration for the benefit of humanity.

## Basic Concept of GPS

## Fundamentals

The GPS concept is based on time and the known position of specialized satellites. The satellites carry very stable atomic clocks that are synchronized with one another and to ground clocks. Any drift from true time maintained on the ground is corrected daily. Likewise, the satellite locations are known with great precision. GPS receivers have clocks as well; however, they are usually not synchronized with true time, and are less stable. GPS satellites continuously transmit their current time and position. A GPS receiver monitors multiple satellites and solves equations to determine the precise position of the receiver and its deviation from true time. At a minimum, four satellites must be in view of the receiver for it to compute four unknown quantities (three position coordinates and clock deviation from satellite time).

## More Detailed Description

Each GPS satellite continually broadcasts a signal (carrier wave with modulation) that includes:

- A pseudorandom code (sequence of ones and zeros) that is known to the receiver. By time-aligning a receiver-generated version and the receiver-measured version of the code, the time of arrival (TOA) of a defined point in the code sequence, called an epoch, can be found in the receiver clock time scale

- A message that includes the time of transmission (TOT) of the code epoch (in GPS system time scale) and the satellite position at that time

Conceptually, the receiver measures the TOAs (according to its own clock) of four satellite signals. From the TOAs and the TOTs, the receiver forms four time of flight (TOF) values, which are (given the speed of light) approximately equivalent to receiver-satellite range differences. The receiver then computes its three-dimensional position and clock deviation from the four TOFs.

In practice the receiver position (in three dimensional Cartesian coordinates with origin at the Earth's center) and the offset of the receiver clock relative to the GPS time are computed simultaneously, using the navigation equations to process the TOFs.

The receiver's Earth-centered solution location is usually converted to latitude, longitude and height relative to an ellipsoidal Earth model. The height may then be further converted to height relative to the geoid (e.g., EGM96) (essentially, mean sea level). These coordinates may be displayed, e.g., on a moving map display, and/or recorded and/or used by some other system (e.g., a vehicle guidance system).

## User-satellite Geometry

Although usually not formed explicitly in the receiver processing, the conceptual time differences of arrival (TDOAs) define the measurement geometry. Each TDOA corresponds to a hyperboloid of revolution. The line connecting the two satellites involved (and its extensions) forms the axis of the hyperboloid. The receiver is located at the point where three hyperboloids intersect.

It is sometimes incorrectly said that the user location is at the intersection of three spheres. While simpler to visualize, this is only the case if the receiver has a clock synchronized with the satellite clocks (i.e., the receiver measures true ranges to the satellites rather than range differences). There are significant performance benefits to the user carrying a clock synchronized with the satellites. Foremost is that only three satellites are needed to compute a position solution. If this were part of the GPS system concept so that all users needed to carry a synchronized clock, then a smaller number of satellites could be deployed. However, the cost and complexity of the user equipment would increase significantly.

## Receiver in Continuous Operation

The description above is representative of a receiver start-up situation. Most receivers have a track algorithm, sometimes called a *tracker*, that combines sets of satellite measurements collected at different times—in effect, taking advantage of the fact that successive receiver positions are usually close to each other. After a set of measurements are processed, the tracker predicts the receiver location corresponding to the next set of satellite measurements. When the new measurements are collected, the receiver uses a weighting scheme to combine the new measurements with the tracker prediction. In general, a tracker can (a) improve receiver position and time accuracy, (b) reject bad measurements, and (c) estimate receiver speed and direction.

The disadvantage of a tracker is that changes in speed or direction can only be computed with a delay, and that derived direction becomes inaccurate when the distance traveled between two position measurements drops below or near the random error of position measurement. GPS units can use measurements of the Doppler shift of the signals received to compute velocity accurately. More advanced navigation systems use additional sensors like a compass or an inertial navigation system to complement GPS.

## Non-navigation Applications

In typical GPS operation as a navigator, four or more satellites must be visible to obtain an accurate result. The solution of the navigation equations gives the position of the receiver along with the difference between the time kept by the receiver's on-board clock and the true time-of-day, thereby eliminating the need for a more precise and possibly impractical receiver based clock. Applications for GPS such as time transfer, traffic signal timing, and synchronization of cell phone base stations, make use of this cheap and highly accurate timing. Some GPS applications use this time for display, or, other than for the basic position calculations, do not use it at all.

Although four satellites are required for normal operation, fewer apply in special cases. If one variable is already known, a receiver can determine its position using only three satellites. For example, a ship or aircraft may have known elevation. Some GPS receivers may use additional clues or assumptions such as reusing the last known altitude, dead reckoning, inertial navigation, or including information from the vehicle computer, to give a (possibly degraded) position when fewer than four satellites are visible.

## Structure

The current GPS consists of three major segments. These are the space segment (SS), a control segment (CS), and a user segment (US). The U.S. Air Force develops, maintains, and operates the space and control segments. GPS satellites broadcast signals from space, and each GPS receiver uses these signals to calculate its three-dimensional location (latitude, longitude, and altitude) and the current time.

The space segment is composed of 24 to 32 satellites in medium Earth orbit and also includes the payload adapters to the boosters required to launch them into orbit. The control segment is composed of a master control station (MCS), an alternate master control station, and a host of dedicated and shared ground antennas and monitor stations. The user segment is composed of hundreds of thousands of U.S. and allied military users of the secure GPS Precise Positioning Service, and hundreds of millions of civil, commercial, and scientific users of the Standard Positioning Service.

## Space Segment

The space segment (SS) is composed of the orbiting GPS satellites, or Space Vehicles (SV) in GPS parlance. The GPS design originally called for 24 SVs, eight each in three approximately circular orbits, but this was modified to six orbital planes with four satellites each. The six orbit planes have approximately 55° inclination (tilt relative to the Earth's equator) and are separated by 60° right ascension of the ascending node (angle along the equator from a reference point to the orbit's intersection). The orbital period is one-half a sidereal day, i.e., 11 hours and 58 minutes so that the satellites pass over the same locations or almost the same locations every day. The orbits are arranged so that at least six satellites are always within line of sight from almost everywhere on the Earth's surface. The result of this objective is that the four satellites are not evenly spaced (90 degrees) apart within each orbit. In general terms, the angular difference between satellites in each orbit is 30, 105, 120, and 105 degrees apart, which sum to 360 degrees.

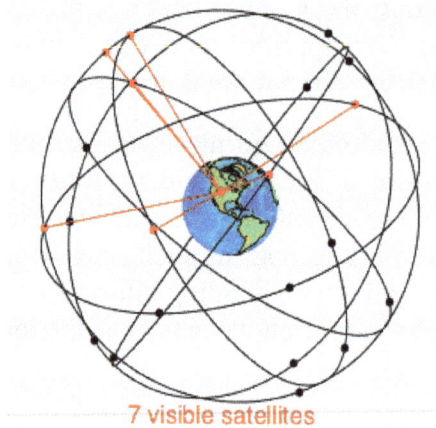

7 visible satellites

A visual example of a 24 satellite GPS constellation in motion with the earth rotating. Notice how the number of *satellites in view* from a given point on the earth's surface, in this example in Golden CO (39.7469° N, 105.2108° W), changes with time.

Orbiting at an altitude of approximately 20,200 km (12,600 mi); orbital radius of approximately 26,600 km (16,500 mi), each SV makes two complete orbits each sidereal day, repeating the same ground track each day. This was very helpful during development because even with only four satellites, correct alignment means all four are visible from one spot for a few hours each day. For military operations, the ground track repeat can be used to ensure good coverage in combat zones.

As of February 2016, there are 32 satellites in the GPS constellation, 31 of which are in use. The additional satellites improve the precision of GPS receiver calculations by providing redundant measurements. With the increased number of satellites, the constellation was changed to a nonuniform arrangement. Such an arrangement was shown to improve reliability and availability of the system, relative to a uniform system, when multiple satellites fail. About nine satellites are visible from any point on the ground at any one time, ensuring considerable redundancy over the minimum four satellites needed for a position.

# Control Segment

Ground monitor station used from 1984 to 2007, on display at
the Air Force Space & Missile Museum.

The control segment is composed of:

1.  a master control station (MCS),

2.  an alternate master control station,

3.  four dedicated ground antennas, and

4.  six dedicated monitor stations.

The MCS can also access U.S. Air Force Satellite Control Network (AFSCN) ground antennas (for additional command and control capability) and NGA (National Geospatial-Intelligence Agency) monitor stations. The flight paths of the satellites are tracked by dedicated U.S. Air Force monitoring stations in Hawaii, Kwajalein Atoll, Ascension Island, Diego Garcia, Colorado Springs, Colorado and Cape Canaveral, along with shared NGA monitor stations operated in England, Argentina, Ecuador, Bahrain, Australia and Washington DC. The tracking information is sent to the Air Force Space Command MCS at Schriever Air Force Base 25 km (16 mi) ESE of Colorado Springs, which is operated by the 2nd Space Operations Squadron (2 SOPS) of the U.S. Air Force. Then 2 SOPS contacts each GPS satellite regularly with a navigational update using dedicated or shared (AFSCN) ground antennas (GPS dedicated ground antennas are located at Kwajalein, Ascension Island, Diego Garcia, and Cape Canaveral). These updates synchronize the atomic clocks on board the satellites to within a few nanoseconds of each other, and adjust the ephemeris of each satellite's internal orbital model. The updates are created by a Kalman filter that uses inputs from the ground monitoring stations, space weather information, and various other inputs.

Satellite maneuvers are not precise by GPS standards—so to change a satellite's orbit, the satellite must be marked *unhealthy*, so receivers don't use it. After the satellite maneuver, engineers track the new orbit from the ground, upload the new ephemeris, and mark the satellite healthy again.

The Operation Control Segment (OCS) currently serves as the control segment of record. It provides the operational capability that supports GPS users and keeps the GPS system operational and performing within specification.

OCS successfully replaced the legacy 1970s-era mainframe computer at Schriever Air Force Base in September 2007. After installation, the system helped enable upgrades and provide a foundation for a new security architecture that supported U.S. armed forces. OCS will continue to be the ground control system of record until the new segment, Next Generation GPS Operation Control System (OCX), is fully developed and functional.

The new capabilities provided by OCX will be the cornerstone for revolutionizing GPS's mission capabilities, and enabling Air Force Space Command to greatly enhance GPS operational services to U.S. combat forces, civil partners and myriad domestic and international users.

The GPS OCX program also will reduce cost, schedule and technical risk. It is designed to provide 50% sustainment cost savings through efficient software architecture and Performance-Based Logistics. In addition, GPS OCX is expected to cost millions less than the cost to upgrade OCS while providing four times the capability.

The GPS OCX program represents a critical part of GPS modernization and provides significant information assurance improvements over the current GPS OCS program.

- OCX will have the ability to control and manage GPS legacy satellites as well as the next generation of GPS III satellites, while enabling the full array of military signals.

- Built on a flexible architecture that can rapidly adapt to the changing needs of today's and future GPS users allowing immediate access to GPS data and constellation status through secure, accurate and reliable information.

- Provides the warfighter with more secure, actionable and predictive information to enhance situational awareness.

- Enables new modernized signals (L1C, L2C, and L5) and has M-code capability, which the legacy system is unable to do.

- Provides significant information assurance improvements over the current program including detecting and preventing cyber attacks, while isolating, containing and operating during such attacks.

- Supports higher volume near real-time command and control capabilities and abilities.

On September 14, 2011, the U.S. Air Force announced the completion of GPS OCX Preliminary Design Review and confirmed that the OCX program is ready for the next phase of development.

The GPS OCX program has missed major milestones and is pushing the GPS IIIA launch beyond April 2016.

# User Segment

GPS receivers come in a variety of formats, from devices
integrated into cars, phones, and watches, to dedicated devices such as these.

The first portable GPS unit, Leica WM 101 displayed at
the Irish National Science Museum at Maynooth.

The user segment is composed of hundreds of thousands of U.S. and allied military users of the secure GPS Precise Positioning Service, and tens of millions of civil, commercial and scientific users of the Standard Positioning Service. In general, GPS receivers are composed of an antenna, tuned to the frequencies transmitted by the satellites, receiver-processors, and a highly stable clock (often a crystal oscillator). They may also include a display for providing location and speed information to the user. A receiver is often described by its number of channels: this signifies how many satellites it can monitor simultaneously. Originally limited to four or five, this has progressively increased over the years so that, as of 2007, receivers typically have between 12 and 20 channels.

A typical OEM GPS receiver module measuring 15×17 mm.

GPS receivers may include an input for differential corrections, using the RTCM SC-104 format. This is typically in the form of an RS-232 port at 4,800 bit/s speed. Data is actually sent at a much lower rate, which limits the accuracy of the signal sent using RTCM. Receivers with internal DGPS receivers can outperform those using external RTCM data. As of 2006, even low-cost units commonly include Wide Area Augmentation System (WAAS) receivers.

A typical GPS receiver with integrated antenna.

Many GPS receivers can relay position data to a PC or other device using the NMEA 0183 protocol. Although this protocol is officially defined by the National Marine Electronics Association (NMEA), references to this protocol have been compiled from public records, allowing open source tools like gpsd to read the protocol without violating intellectual property laws. Other proprietary protocols exist as well, such as the SiRF and MTK protocols. Receivers can interface with other devices using methods including a serial connection, USB, or Bluetooth.

## Applications

While originally a military project, GPS is considered a *dual-use* technology, meaning it has significant military and civilian applications.

GPS has become a widely deployed and useful tool for commerce, scientific uses, tracking, and surveillance. GPS's accurate time facilitates everyday activities such as banking, mobile phone operations, and even the control of power grids by allowing well synchronized hand-off switching.

## Civilian

This antenna is mounted on the roof of a hut
containing a scientific experiment needing precise timing.

Many civilian applications use one or more of GPS's three basic components: absolute location, relative movement, and time transfer.

- Astronomy: both positional and clock synchronization data is used in astrometry and celestial mechanics. GPS is also used in both amateur astronomy with small telescopes as well as by professional observatories for finding extrasolar planets, for example.

- Automated vehicle: applying location and routes for cars and trucks to function without a human driver.

- Cartography: both civilian and military cartographers use GPS extensively.

- Cellular telephony: clock synchronization enables time transfer, which is critical for synchronizing its spreading codes with other base stations to facilitate inter-cell handoff and support hybrid GPS/cellular position detection for mobile emergency calls and other applications. The first handsets with integrated GPS launched in the late 1990s. The U.S. Federal Communications Commission (FCC) mandated the feature in either the handset or in the towers (for use in triangulation) in 2002 so emergency services could locate 911 callers. Third-party software developers later gained access to GPS APIs from Nextel upon launch, followed by Sprint in 2006, and Verizon soon thereafter.

- Clock synchronization: the accuracy of GPS time signals (±10 ns) is second only to the atomic clocks they are based on.

- Disaster relief/emergency services: many emergency services depend upon GPS for location and timing capabilities.

- GPS-equipped radiosondes and dropsondes: measure and calculate the atmospheric pressure, wind speed and direction up to 27 km from the Earth's surface.

- Radio occultation for weather and atmospheric science applications.

- Fleet tracking: used to identify, locate and maintain contact reports with one or more fleet vehicles in real-time.

- Geofencing: vehicle tracking systems, person tracking systems, and pet tracking systems use GPS to locate devices that are attached to or carried by a person, vehicle, or pet. The application can provide continuous tracking and send notifications if the target leaves a designated (or "fenced-in") area.

- Geotagging: applies location coordinates to digital objects such as photographs (in Exif data) and other documents for purposes such as creating map overlays with devices like Nikon GP-1

- GPS aircraft tracking

- GPS for mining: the use of RTK GPS has significantly improved several mining operations such as drilling, shoveling, vehicle tracking, and surveying. RTK GPS provides centimeter-level positioning accuracy.

- GPS data mining: It is possible to aggregate GPS data from multiple users to understand movement patterns, common trajectories and interesting locations.

- GPS tours: location determines what content to display; for instance, information about an approaching point of interest.

- Navigation: navigators value digitally precise velocity and orientation measurements.

- Phasor measurements: GPS enables highly accurate timestamping of power system measurements, making it possible to compute phasors.

- Recreation: for example, Geocaching, Geodashing, GPS drawing, waymarking, and other kinds of location based mobile games.

- Robotics: self-navigating, autonomous robots using a GPS sensors, which calculate latitude, longitude, time, speed, and heading.

- Sport: used in football and rugby for the control and analysis of the training load.

- Surveying: surveyors use absolute locations to make maps and determine property boundaries.

- Tectonics: GPS enables direct fault motion measurement of earthquakes. Between earthquakes GPS can be used to measure crustal motion and deformation to estimate seismic strain buildup for creating seismic hazard maps.

- Telematics: GPS technology integrated with computers and mobile communications technology in automotive navigation systems.

## Restrictions on Civilian Use

The U.S. government controls the export of some civilian receivers. All GPS receivers capable of functioning above 18 km (60,000 feet) altitude and 515 m/s (1,000 knots), or designed or modified for use with unmanned air vehicles like, e.g., ballistic or cruise missile systems, are classified as munitions (weapons)—which means they require State Department export licenses.

This rule applies even to otherwise purely civilian units that only receive the L1 frequency and the C/A (Coarse/Acquisition) code.

Disabling operation above these limits exempts the receiver from classification as a munition. Vendor interpretations differ. The rule refers to operation at both the target altitude and speed, but some receivers stop operating even when stationary. This has caused problems with some amateur radio balloon launches that regularly reach 30 km (100,000 feet).

These limits only apply to units or components exported from the USA. A growing trade in various components exists, including GPS units from other countries. These are expressly sold as ITAR-free.

## Military

Attaching a GPS guidance kit to a dumb bomb, March 2003.

M982 Excalibur GPS-guided artillery shell.

As of 2009, military GPS applications include:

- Navigation: Soldiers use GPS to find objectives, even in the dark or in unfamiliar territory, and to coordinate troop and supply movement. In the United States armed forces, commanders use the *Commanders Digital Assistant* and lower ranks use the *Soldier Digital Assistant.*

- Target tracking: Various military weapons systems use GPS to track potential ground and air targets before flagging them as hostile. These weapon systems pass target coordinates to precision-guided munitions to allow them to engage targets accurately. Military aircraft, particularly in air-to-ground roles, use GPS to find targets.

- Missile and projectile guidance: GPS allows accurate targeting of various military weapons including ICBMs, cruise missiles, precision-guided munitions and Artillery projectiles. Embedded GPS receivers able to withstand accelerations of 12,000 $g$ or about 118 km/s$^2$ have been developed for use in 155-millimeter (6.1 in) howitzer shells.

- Search and rescue.

- Reconnaissance: Patrol movement can be managed more closely.

- GPS satellites carry a set of nuclear detonation detectors consisting of an optical sensor (Y-sensor), an X-ray sensor, a dosimeter, and an electromagnetic pulse (EMP) sensor (W-sensor), that form a major portion of the United States Nuclear Detonation Detection System. General William Shelton has stated that future satellites may drop this feature to save money.

GPS type navigation was first used in war in the 1991 Persian Gulf War, before GPS was fully developed in 1995, to assist Coalition Forces to navigate and perform maneuvers in the war. The war also demonstrated the vulnerability of GPS to being jammed, when Iraqi forces added noise to the weak GPS signal transmission to protect Iraqi targets.

## Communication

The navigational signals transmitted by GPS satellites encode a variety of information including satellite positions, the state of the internal clocks, and the health of the network. These signals are transmitted on two separate carrier frequencies that are common to all satellites in the network. Two different encodings are used: a public encoding that enables lower resolution navigation, and an encrypted encoding used by the U.S. military.

## Message Format

| GPS message format | |
| --- | --- |
| **Subframes** | **Description** |
| 1 | Satellite clock, GPS time relationship |
| 2–3 | Ephemeris (precise satellite orbit) |
| 4–5 | Almanac component (satellite network synopsis, error correction) |

Each GPS satellite continuously broadcasts a *navigation message* on L1 C/A and L2 P/Y frequencies at a rate of 50 bits per second. Each complete message takes 750 seconds (12 1/2 minutes) to complete. The message structure has a basic format of a 1500-bit-long frame made up of five subframes, each subframe being 300 bits (6 seconds) long. Subframes 4 and 5 are subcommutated 25 times each, so that a complete data message requires the transmission of 25 full frames. Each subframe consists of ten words, each 30 bits long. Thus, with 300 bits in a subframe times 5 subframes in a frame times 25 frames in a message, each message is 37,500 bits long. At a transmission rate of 50-bit/s, this gives 750 seconds to transmit an entire almanac message (GPS). Each 30-second frame begins precisely on the minute or half-minute as indicated by the atomic clock on each satellite.

The first subframe of each frame encodes the week number and the time within the week, as well as the data about the health of the satellite. The second and the third subframes contain the *ephemeris* – the precise orbit for the satellite. The fourth and fifth subframes contain the *almanac*, which contains coarse orbit and status information for up to 32 satellites in the constellation as well as data related to error correction. Thus, to obtain an accurate satellite location from this transmitted message, the receiver must demodulate the message from each satellite it includes in its solution for 18 to 30 seconds. To collect all transmitted almanacs, the receiver must demodulate the message for 732 to 750 seconds or 12 1/2 minutes.

All satellites broadcast at the same frequencies, encoding signals using unique code division multiple access (CDMA) so receivers can distinguish individual satellites from each other. The system uses two distinct CDMA encoding types: the coarse/acquisition (C/A) code, which is accessible by the general public, and the precise (P(Y)) code, which is encrypted so that only the U.S. military and other NATO nations who have been given access to the encryption code can access it.

The ephemeris is updated every 2 hours and is generally valid for 4 hours, with provisions for updates every 6 hours or longer in non-nominal conditions. The almanac is updated typically every 24 hours. Additionally, data for a few weeks following is uploaded in case of transmission updates that delay data upload.

## Satellite Frequencies

| GPS frequency overview | | |
|---|---|---|
| **Band** | **Frequency** | **Description** |
| L1 | 1575.42 MHz | Coarse-acquisition (C/A) and encrypted precision (P(Y)) codes, plus the L1 civilian (L1C) and military (M) codes on future Block III satellites. |
| L2 | 1227.60 MHz | P(Y) code, plus the L2C and military codes on the Block IIR-M and newer satellites. |
| L3 | 1381.05 MHz | Used for nuclear detonation (NUDET) detection. |
| L4 | 1379.913 MHz | Being studied for additional ionospheric correction. |
| L5 | 1176.45 MHz | Proposed for use as a civilian safety-of-life (SoL) signal. |

All satellites broadcast at the same two frequencies, 1.57542 GHz (L1 signal) and 1.2276 GHz (L2 signal). The satellite network uses a CDMA spread-spectrum technique where the low-bitrate message data is encoded with a high-rate pseudo-random (PRN) sequence that is different for each satellite. The receiver must be aware of the PRN codes for each satellite to reconstruct the actual message data. The C/A code, for civilian use, transmits data at 1.023 million chips per second, whereas the P code, for U.S. military use, transmits at 10.23 million chips per second. The actual internal reference of the satellites is 10.22999999543 MHz to compensate for relativistic effects that make observers on the Earth perceive a different time reference with respect to the transmitters in orbit. The L1 carrier is modulated by both the C/A and P codes, while the L2 carrier is only modulated by the P code. The P code can be encrypted as a so-called P(Y) code that is only available to military equipment with a proper decryption key. Both the C/A and P(Y) codes impart the precise time-of-day to the user.

The L3 signal at a frequency of 1.38105 GHz is used to transmit data from the satellites to ground stations. This data is used by the United States Nuclear Detonation (NUDET) Detection System (USNDS) to detect, locate, and report nuclear detonations (NUDETs) in the Earth's atmosphere and near space. One usage is the enforcement of nuclear test ban treaties.

The L4 band at 1.379913 GHz is being studied for additional ionospheric correction.

The L5 frequency band at 1.17645 GHz was added in the process of GPS modernization. This frequency falls into an internationally protected range for aeronautical navigation, promising little or no interference under all circumstances. The first Block IIF satellite that provides this signal was launched in 2010. The L5 consists of two carrier components that are in phase quadrature with each other. Each carrier component is bi-phase shift key (BPSK) modulated by a separate bit train. "L5, the third civil GPS signal, will eventually support safety-of-life applications for aviation and provide improved availability and accuracy."

A conditional waiver has recently (2011-01-26) been granted to LightSquared to operate a terrestrial broadband service near the L1 band. Although LightSquared had applied for a license to operate in the 1525 to 1559 band as early as 2003 and it was put out for public comment, the FCC asked LightSquared to form a study group with the GPS community to test GPS receivers and identify issue that might arise due to the larger signal power from the LightSquared terrestrial network. The GPS community had not objected to the LightSquared (formerly MSV and SkyTerra) applications until November 2010, when LightSquared applied for a modification to its Ancillary Terrestrial Component (ATC) authorization. This filing (SAT-MOD-20101118-00239) amounted to a request to run several orders of magnitude more power in the same frequency band for terrestrial base stations, essentially repurposing what was supposed to be a "quiet neighborhood" for signals from space as the equivalent of a cellular network. Testing in the first half of 2011 has demonstrated that the impact of the lower 10 MHz of spectrum is minimal to GPS devices (less than 1% of the total GPS devices are affected). The upper 10 MHz intended for use by LightSquared may have some impact on GPS devices. There is some concern that this may seriously degrade the GPS signal for many consumer uses. Aviation Week magazine reports that the latest testing (June 2011) confirms "significant jamming" of GPS by LightSquared's system.

## Demodulation and Decoding

Because all of the satellite signals are modulated onto the same L1 carrier frequency, the signals must be separated after demodulation. This is done by assigning each satellite a unique binary sequence known as a Gold code. The signals are decoded after demodulation using addition of the Gold codes corresponding to the satellites monitored by the receiver.

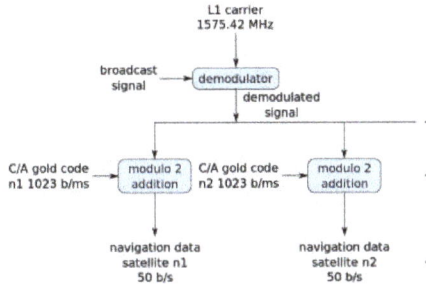

Demodulating and Decoding GPS Satellite Signals using the Coarse/Acquisition Gold code.

If the almanac information has previously been acquired, the receiver picks the satellites to listen for by their PRNs, unique numbers in the range 1 through 32. If the almanac information is not in memory, the receiver enters a search mode until a lock is obtained on one of the satellites. To obtain a lock, it is necessary that there be an unobstructed line of sight from the receiver to the satellite. The receiver can then acquire the almanac and determine the satellites it should listen for. As it detects each satellite's signal, it identifies it by its distinct C/A code pattern. There can be a delay of up to 30 seconds before the first estimate of position because of the need to read the ephemeris data.

Processing of the navigation message enables the determination of the time of transmission and the satellite position at this time.

## Navigation Equations

## Problem Description

The receiver uses messages received from satellites to determine the satellite positions and time sent. The $x$, $y$, and $z$ components of satellite position and the time sent are designated as $[x_i, y_i, z_i, s_i]$ where the subscript $i$ denotes the satellite and has the value 1, 2, ..., $n$, where $n \geq 4$. When the time of message reception indicated by the on-board receiver clock is $\tilde{t}_i$, the true reception time is $t_i = \tilde{t}_i - b$, where $b$ is the receiver's clock bias from the much more accurate GPS system clocks employed by the satellites. The receiver clock bias is the same for all received satellite signals (assuming the satellite clocks are all perfectly synchronized). The message's transit time is $\tilde{t}_i - b - s_i$, where $s_i$ is the satellite time. Assuming the message traveled at the speed of light, $c$, the distance traveled is $(\tilde{t}_i - b - s_i)c$.

For n satellites, the equations to satisfy are:

$$(x - x_i)^2 + (y - y_i)^2 + (z - z_i)^2 = ([\tilde{t}_i - b - s_i]c)^2, i = 1, 2, \ldots, n$$

or in terms of pseudoranges, $p_i = (\tilde{t}_i - s_i)c$, as

$$\sqrt{(x-x_i)^2 + (y-y_i)^2 + (z-z_i)^2} + bc = p_i, i = 1, 2, ..., n.$$

Since the equations have four unknowns [$x$, $y$, $z$, $b$]—the three components of GPS receiver position and the clock bias—signals from at least four satellites are necessary to attempt solving these equations. They can be solved by algebraic or numerical methods. Existence and uniqueness of GPS solutions are discussed by Abell and Chaffee. When $n$ is greater than 4 this system is overdetermined and a fitting method must be used.

With each combination of satellites, GDOP quantities can be calculated based on the relative sky directions of the satellites used. The receiver location is expressed in a specific coordinate system, such as latitude and longitude using the WGS 84 geodetic datum or a country-specific system.

## Geometric Interpretation

The GPS equations can be solved by numerical and analytical methods. Geometrical interpretations can enhance the understanding of these solution methods.

## Spheres

The measured ranges, called pseudoranges, contain clock errors. In a simplified idealization in which the ranges are synchronized, these true ranges represent the radii of spheres, each centered on one of the transmitting satellites. The solution for the position of the receiver is then at the intersection of the surfaces of three of these spheres. If more than the minimum number of ranges is available, a near intersection of more than three sphere surfaces could be found via, e.g. least squares.

## Hyperboloids

If the distance traveled between the receiver and satellite $i$ and the distance traveled between the receiver and satellite $j$ are subtracted, the result is $(\tilde{t}_i - s_i)c - (\tilde{t}_j - s_j)c$, which only involves known or measured quantities. The locus of points having a constant difference in distance to two points (here, two satellites) is a hyperboloid. Thus, from four or more measured reception times, the receiver can be placed at the intersection of the surfaces of three or more hyperboloids.

## Spherical Cones

The solution space [$x$, $y$, $z$, $b$] can be seen as a four-dimensional geometric space. In that case each of the equations describes a spherical cone, with the cusp located at the satellite, and the base a sphere around the satellite. The receiver is at the intersection of four or more of such cones.

## Solution Methods

## Least Squares

When more than four satellites are available, the calculation can use the four best, or more than four simultaneously (up to all visible satellites), depending on the number of receiver channels, processing capability, and geometric dilution of precision (GDOP).

Using more than four involves an over-determined system of equations with no unique solution; such a system can be solved by a least-squares or weighted least squares method.

$$\left(\hat{x}, \hat{y}, \hat{z}, \hat{b}\right) = \underset{(x,y,z,b)}{\arg\min} \sum_i \left(\sqrt{(x-x_i)^2 + (y-y_i)^2 + (z-z_i)^2} + bc - p_i\right)^2$$

## Iterative

Both the equations for four satellites, or the least squares equations for more than four, are non-linear and need special solution methods. A common approach is by iteration on a linearized form of the equations, such as the Gauss–Newton algorithm.

The GPS system was initially developed assuming use of a numerical least-squares solution method—i.e., before closed-form solutions were found.

## Closed-form

One closed-form solution to the above set of equations was developed by S. Bancroft. Its properties are well known; in particular, proponents claim it is superior in low-GDOP situations, compared to iterative least squares methods.

Bancroft's method is algebraic, as opposed to numerical, and can be used for four or more satellites. When four satellites are used, the key steps are inversion of a 4x4 matrix and solution of a single-variable quadratic equation. Bancroft's method provides one or two solutions for the unknown quantities. When there are two (usually the case), only one is a near-Earth sensible solution.

When a receiver uses more than four satellites for a solution, Bancroft uses the generalized inverse (i.e., the pseudoinverse) to find a solution. However, a case has been made that iterative methods (e.g., Gauss–Newton algorithm) for solving over-determined non-linear least squares (NLLS) problems generally provide more accurate solutions.

Leick et al. (2015) states that "Bancroft's (1985) solution is a very early, if not the first, closed-form solution." Other closed-form solutions were published afterwards, although their adoption in practice is unclear.

## Error Sources and Analysis

GPS error analysis examines error sources in GPS results and the expected size of those errors. GPS makes corrections for receiver clock errors and other effects, but some residual errors remain uncorrected. Error sources include signal arrival time measurements, numerical calculations, atmospheric effects (ionospheric/ tropospheric delays), ephemeris and clock data, multipath signals, and natural and artificial interference. Magnitude of residual errors from these sources depends on geometric dilution of precision. Artificial errors may result from jamming devices and threaten ships and aircraft or from intentional signal degradation through selective availability, which limited accuracy to ~6–12 m, but has been switched off since May 1, 2000.

## Accuracy Enhancement and Surveying

## Augmentation

Integrating external information into the calculation process can materially improve accuracy. Such augmentation systems are generally named or described based on how the information arrives. Some systems transmit additional error information (such as clock drift, ephemera, or ionospheric delay), others characterize prior errors, while a third group provides additional navigational or vehicle information.

Examples of augmentation systems include the Wide Area Augmentation System (WAAS), European Geostationary Navigation Overlay Service (EGNOS), Differential GPS (DGPS), inertial navigation systems (INS) and Assisted GPS. The standard accuracy of about 15 meters (49 feet) can be augmented to 3–5 meters (9.8–16.4 ft) with DGPS, and to about 3 meters (9.8 feet) with WAAS.

## Precise Monitoring

Accuracy can be improved through precise monitoring and measurement of existing GPS signals in additional or alternate ways.

The largest remaining error is usually the unpredictable delay through the ionosphere. The spacecraft broadcast ionospheric model parameters, but some errors remain. This is one reason GPS spacecraft transmit on at least two frequencies, L1 and L2. Ionospheric delay is a well-defined function of frequency and the total electron content (TEC) along the path, so measuring the arrival time difference between the frequencies determines TEC and thus the precise ionospheric delay at each frequency.

Military receivers can decode the P(Y) code transmitted on both L1 and L2. Without decryption keys, it is still possible to use a *codeless* technique to compare the P(Y) codes on L1 and L2 to gain much of the same error information. However, this technique is slow, so it is currently available only on specialized surveying equipment. In the future, additional civilian codes are expected to be transmitted on the L2 and L5 frequencies. All users will then be able to perform dual-frequency measurements and directly compute ionospheric delay errors.

A second form of precise monitoring is called *Carrier-Phase Enhancement* (CPGPS). This corrects the error that arises because the pulse transition of the PRN is not instantaneous, and thus the correlation (satellite–receiver sequence matching) operation is imperfect. CPGPS uses the L1 carrier wave, which has a period of

$$\frac{1s}{1575.42 \times 10^6} = 0.63475ns \approx 1ns$$ , which is about one-thousandth of the C/A Gold code bit

period of $$\frac{1s}{1023 \times 10^3} = 977.5ns \approx 1000ns$$ , to act as an additional clock signal and resolve

the uncertainty. The phase difference error in the normal GPS amounts to 2–3 meters (7–10 ft) of ambiguity. CPGPS working to within 1% of perfect transition reduces this error to 3 centimeters (1.2 in) of ambiguity. By eliminating this error source, CPGPS coupled with DGPS normally realizes between 20–30 centimeters (8–12 in) of absolute accuracy.

*Relative Kinematic Positioning* (RKP) is a third alternative for a precise GPS-based positioning system. In this approach, determination of range signal can be resolved to a precision of less than 10 centimeters (4 in). This is done by resolving the number of cycles that the signal is transmitted and received by the receiver by using a combination of differential GPS (DGPS) correction data, transmitting GPS signal phase information and ambiguity resolution techniques via statistical tests—possibly with processing in real-time (real-time kinematic positioning, RTK).

## Timekeeping

## Leap Seconds

While most clocks derive their time from Coordinated Universal Time (UTC), the atom-ic clocks on the satellites are set to GPS time. The difference is that GPS time is not corrected to match the rotation of the Earth, so it does not contain leap seconds or other corrections that are periodically added to UTC. GPS time was set to match UTC in 1980, but has since diverged. The lack of corrections means that GPS time remains at a constant offset with International Atomic Time (TAI) (TAI – GPS = 19 seconds). Periodic corrections are performed to the on-board clocks to keep them synchronized with ground clocks.

The GPS navigation message includes the difference between GPS time and UTC. As of July 2015, GPS time is 17 seconds ahead of UTC because of the leap second added to UTC on June 30, 2015. Receivers subtract this offset from GPS time to calculate UTC and specific timezone values. New GPS units may not show the correct UTC time until after receiving the UTC offset message. The GPS-UTC offset field can accommodate 255 leap seconds (eight bits).

## Accuracy

GPS time is theoretically accurate to about 14 nanoseconds. However, most receivers lose accuracy in the interpretation of the signals and are only accurate to 100 nanoseconds.

## Format

As opposed to the year, month, and day format of the Gregorian calendar, the GPS date is expressed as a week number and a seconds-into-week number. The week number is transmitted as a ten-bit field in the C/A and P(Y) navigation messages, and so it becomes zero again every 1,024 weeks (19.6 years). GPS week zero started at 00:00:00 UTC (00:00:19 TAI) on January 6, 1980, and the week number became zero again for the first time at 23:59:47 UTC on August 21, 1999 (00:00:19 TAI on August 22, 1999). To determine the current Gregorian date, a GPS receiver must be provided with the approximate date (to within 3,584 days) to correctly translate the GPS date signal. To address this concern the modernized GPS navigation message uses a 13-bit field that only repeats every 8,192 weeks (157 years), thus lasting until the year 2137 (157 years after GPS week zero).

## Carrier Phase Tracking (Surveying)

Another method that is used in surveying applications is carrier phase tracking. The period of the carrier frequency multiplied by the speed of light gives the wavelength, which is about 0.19 meters for the L1 carrier. Accuracy within 1% of wavelength in detecting the leading edge reduces this component of pseudorange error to as little as 2 millimeters. This compares to 3 meters for the C/A code and 0.3 meters for the P code.

However, 2 millimeter accuracy requires measuring the total phase—the number of waves multiplied by the wavelength plus the fractional wavelength, which requires specially equipped receivers. This method has many surveying applications. It is accurate enough for real-time tracking of the very slow motions of tectonic plates, typically 0–100 mm (0–4 inches) per year.

Triple differencing followed by numerical root finding, and a mathematical technique called least squares can estimate the position of one receiver given the position of another. First, compute the difference between satellites, then between receivers, and finally between epochs. Other orders of taking differences are equally valid. Detailed discussion of the errors is omitted.

The satellite carrier total phase can be measured with ambiguity as to the number of cycles. Let $\phi(r_i, s_j, t_k)$ denote the phase of the carrier of satellite $j$ measured by receiver $i$ at time $t_k$. This notation shows the meaning of the subscripts $i, j$, and $k$. The receiver ($r$), satellite ($s$), and time ($t$) come in alphabetical order as arguments of $\phi$ and to balance readability and conciseness, let $\phi_{i,j,k} = \phi(r_i, s_j, t_k)$ be a concise abbreviation. Also we define three functions, : $\Delta^r, \Delta^s, \Delta^t$, , which return differences between receivers, satellites, and time points, respectively. Each function has variables with three subscripts as its arguments. These three functions are defined below. If $\alpha_{i,j,k}$ is a function of the three integer arguments, $i, j,$ and $k$ then it is a valid argument for the functions, : $\Delta^r, \Delta^s, \Delta^t$, with the values defined as

$$\Delta^r(\alpha_{i,j,k}) = \alpha_{i+1,j,k} - \alpha_{i,j,k},$$

$$\Delta^s(\alpha_{i,j,k}) = \alpha_{i,j+1,k} - \alpha_{i,j,k}, \text{ and}$$

$$\Delta^t(\alpha_{i,j,k}) = \alpha_{i,j,k+1} - \alpha_{i,j,k}.$$

Also if $\alpha_{i,j,k}$ *and* $\beta_{l,m,n}$ are valid arguments for the three functions and $a$ and $b$ are constants then $(a\,\alpha_{i,j,k} + b\,\beta_{l,m,n})$ is a valid argument with values defined as

$$\Delta^r(a\,\alpha_{i,j,k} + b\,\beta_{l,m,n}) = a\,\Delta^r(\alpha_{i,j,k}) + b\,\Delta^r(\beta_{l,m,n}),$$

$$\Delta^s(a\,\alpha_{i,j,k} + b\,\beta_{l,m,n}) = a\,\Delta^s(\alpha_{i,j,k}) + b\,\Delta^s(\beta_{l,m,n}), \text{ and}$$

$$\Delta^t(a\,\alpha_{i,j,k} + b\,\beta_{l,m,n}) = a\,\Delta^t(\alpha_{i,j,k}) + b\,\Delta^t(\beta_{l,m,n}).$$

Receiver clock errors can be approximately eliminated by differencing the phases measured from satellite 1 with that from satellite 2 at the same epoch. This difference is designated as $\Delta^s(\phi_{1,1,1}) = \phi_{1,2,1} - \phi_{1,1,1}$

Double differencing computes the difference of receiver 1's satellite difference from that of receiver 2. This approximately eliminates satellite clock errors. This double difference is:

$$\Delta^r(\Delta^s(\phi_{1,1,1})) \quad = \Delta^r(\phi_{1,2,1} - \phi_{1,1,1}) \quad = \Delta^r(\phi_{1,2,1}) - \Delta^r(\phi_{1,1,1}) \quad = (\phi_{2,2,1} - \phi_{1,2,1}) - (\phi_{2,1,1} - \phi_{1,1,1})$$

Triple differencing subtracts the receiver difference from time 1 from that of time 2. This eliminates the ambiguity associated with the integral number of wavelengths in carrier phase provided this ambiguity does not change with time. Thus the triple difference result eliminates practically all clock bias errors and the integer ambiguity. Atmospheric delay and satellite ephemeris errors have been significantly reduced. This triple difference is:

$$\Delta^t(\Delta^r(\Delta^s(\phi_{1,1,1})))$$

Triple difference results can be used to estimate unknown variables. For example, if the position of receiver 1 is known but the position of receiver 2 unknown, it may be possible to estimate the position of receiver 2 using numerical root finding and least squares. Triple difference results for three independent time pairs may be sufficient to solve for receiver 2's three position components. This may require a numerical procedure. An approximation of receiver 2's position is required to use such a numerical method. This initial value can probably be provided from the navigation message and the intersection of sphere surfaces. Such a reasonable estimate can be key to successful multidimensional root finding. Iterating from three time pairs and a fairly good initial value produces one observed triple difference result for receiver 2's position. Processing additional time pairs can improve accuracy, overdetermining the answer with multiple solutions. Least squares can estimate an overdetermined system. Least squares determines the position of receiver 2 that best fits the observed triple difference results for receiver 2 positions under the criterion of minimizing the sum of the squares.

## Regulatory Spectrum Issues Concerning GPS Receivers

In the United States, GPS receivers are regulated under the Federal Communications Commission's (FCC) Part 15 rules. As indicated in the manuals of GPS-enabled devices sold in the United States, as a Part 15 device, it "must accept any interference received, including interference that may cause undesired operation." With respect to GPS devices in particular, the FCC states that GPS receiver manufacturers, "must use receivers that reasonably discriminate against reception of signals outside their allocated spectrum." For the last 30 years, GPS receivers have operated next to the Mobile Satellite Service band, and have discriminated against reception of mobile satellite services, such as Inmarsat, without any issue.

The spectrum allocated for GPS L1 use by the FCC is 1559 to 1610 MHz, while the spectrum allocated for satellite-to-ground use owned by Lightsquared is the Mobile Satellite Service band. Since 1996, the FCC has authorized licensed use of the spectrum neighboring the GPS band of 1525 to 1559 MHz to the Virginia company LightSquared. On March 1, 2001, the FCC received an application from LightSquared's predecessor, Motient Services to use their allocated frequencies for an integrated satellite-terrestrial service. In 2002, the U.S. GPS Industry Council came to an out-of-band-emissions (OOBE) agreement with LightSquared to prevent transmissions from LightSquared's ground-based stations from emitting transmissions into the neighboring GPS band of 1559 to 1610 MHz. In 2004, the FCC adopted the OOBE agreement in its authorization for LightSquared to deploy a ground-based network ancillary to their satellite system – known as the Ancillary Tower Components (ATCs) – "We will authorize MSS ATC subject to conditions that ensure that the added terrestrial component remains ancillary to the principal MSS offering. We do not intend, nor will we permit, the terrestrial component to become a stand-alone service." This authorization was reviewed and approved by the U.S. Interdepartment Radio Advisory Committee, which includes the U.S. Department of Agriculture, U.S. Air Force, U.S. Army, U.S. Coast Guard, Federal Aviation Administration, National Aeronautics and Space Administration, Interior, and U.S. Department of Transportation.

In January 2011, the FCC conditionally authorized LightSquared's wholesale customers—such as Best Buy, Sharp, and C Spire—to only purchase an integrated satellite-ground-based service from LightSquared and re-sell that integrated service on devices that are equipped to only use the ground-based signal using LightSquared's allocated frequencies of 1525 to 1559 MHz. In December 2010, GPS receiver manufacturers expressed concerns to the FCC that LightSquared's signal would interfere with GPS receiver devices although the FCC's policy considerations leading up to the January 2011 order did not pertain to any proposed changes to the maximum number of ground-based LightSquared stations or the maximum power at which these stations could operate. The January 2011 order makes final authorization contingent upon studies of GPS interference issues carried out by a LightSquared led working group along with GPS industry and Federal agency participation. On February 14, 2012, the FCC initiated proceedings to vacate LightSquared's Conditional Waiver Order based on the NTIA's conclusion that there was currently no practical way to mitigate potential GPS interference.

GPS receiver manufacturers design GPS receivers to use spectrum beyond the GPS-allocated band. In some cases, GPS receivers are designed to use up to 400 MHz of spectrum in either direction of the L1 frequency of 1575.42 MHz, because mobile satellite services in those regions are broadcasting from space to ground, and at power levels commensurate with mobile satellite services. However, as regulated under the FCC's Part 15 rules, GPS receivers are not warranted protection from signals outside GPS-allocated spectrum. This is why GPS operates next to the Mobile Satellite Service band, and also why the Mobile Satellite Service band operates next to GPS. The symbiotic relationship of spectrum allocation ensures that users of both bands are able to operate cooperatively and freely.

The FCC adopted rules in February 2003 that allowed Mobile Satellite Service (MSS) licensees such as LightSquared to construct a small number of ancillary ground-based towers in their licensed spectrum to "promote more efficient use of terrestrial wireless spectrum." In those 2003 rules, the FCC stated "As a preliminary matter, terrestrial [Commercial Mobile Radio Service ("CMRS")] and MSS ATC are expected to have different prices, coverage, product acceptance and distribution; therefore, the two services appear, at best, to be imperfect substitutes for one another that would be operating in predominately different market segments... MSS ATC is unlikely to compete directly with terrestrial CMRS for the same customer base...". In 2004, the FCC clarified that the ground-based towers would be ancillary, noting that "We will authorize MSS ATC subject to conditions that ensure that the added terrestrial component remains ancillary to the principal MSS offering. We do not intend, nor will we permit, the terrestrial component to become a stand-alone service." In July 2010, the FCC stated that it expected LightSquared to use its authority to offer an integrated satellite-terrestrial service to "provide mobile broadband services similar to those provided by terrestrial mobile providers and enhance competition in the mobile broadband sector." However, GPS receiver manufacturers have argued that LightSquared's licensed spectrum of 1525 to 1559 MHz was never envisioned as being used for high-speed wireless broadband based on the 2003 and 2004 FCC ATC rulings making clear that the Ancillary Tower Component (ATC) would be, in fact, ancillary to the primary satellite component. To build public support of efforts to continue the 2004 FCC authorization of LightSquared's ancillary terrestrial component vs. a simple ground-based LTE service in the Mobile Satellite Service band, GPS receiver manufacturer Trimble Navigation Ltd. formed the "Coalition To Save Our GPS."

The FCC and LightSquared have each made public commitments to solve the GPS interference issue before the network is allowed to operate. However, according to Chris Dancy of the Aircraft Owners and Pilots Association, airline pilots with the type of systems that would be affected "may go off course and not even realize it." The problems could also affect the Federal Aviation Administration upgrade to the air traffic control system, United States Defense Department guidance, and local emergency services including 911.

On February 14, 2012, the U.S. Federal Communications Commission (FCC) moved to bar LightSquared's planned national broadband network after being informed by the National Telecommunications and Information Administration (NTIA), the federal agency that coordinates spectrum uses for the military and other federal government entities, that "there is no practical way to mitigate potential interference at this time". LightSquared is challenging the FCC's action.

## Other Systems

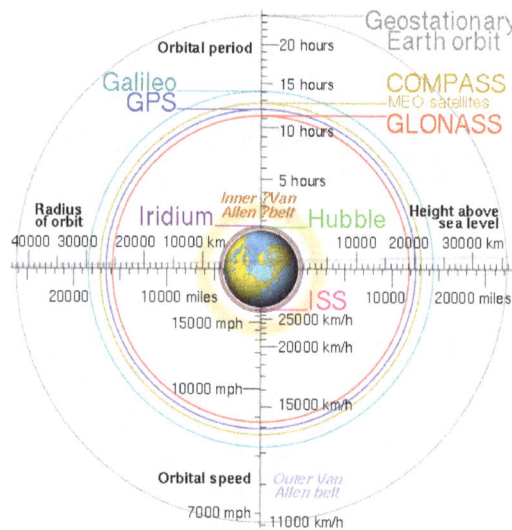

Comparison of geostationary, GPS, GLONASS, Galileo, Compass (MEO), International Space Station, Hubble Space Telescope and Iridium constellation orbits, with the Van Allen radiation belts and the Earth to scale. The Moon's orbit is around 9 times larger than geostationary orbit. (In the SVG file, hover over an orbit or its label to highlight it.)

Other satellite navigation systems in use or various states of development include:

- GLONASS – Russia's global navigation system. Fully operational worldwide.

- Galileo – a global system being developed by the European Union and other partner countries, planned to be operational by 2016 (and fully deployed by 2020)

- Beidou – People's Republic of China's regional system, currently limited to Asia and the West Pacific, global coverage planned to be operational by 2020

- IRNSS (NAVIC) – India's regional navigation system, covering India and Northern Indian Ocean

- QZSS – Japanese regional system covering Asia and Oceania

## GPS Satellite Blocks

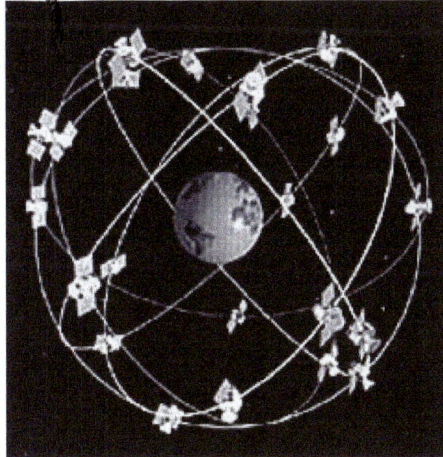

GPS satellite constellation (not to scale)

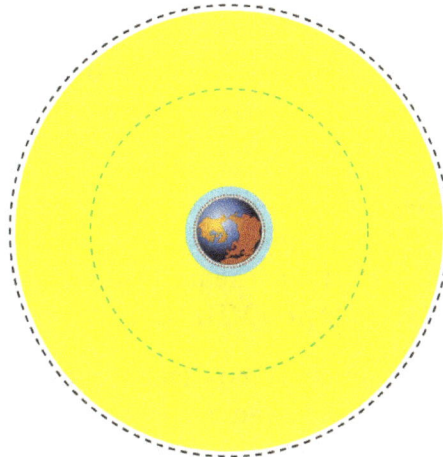

Various earth orbits to scale; green dash-dot line is medium earth orbit, a typical GPS orbit.

A GPS satellite is a satellite used by the NAVSTAR Global Positioning System (GPS). The first satellite in the system, Navstar 1, was launched February 22, 1978. The GPS satellite constellation is operated by the 50th Space Wing of the United States Air Force.

The GPS satellites circle the Earth at an altitude of about 20,000 km (12,427 miles) and complete two full orbits every day.

## Block I Satellites

A Navstar GPS satellite undergoing pre-launch testing.

Rockwell International was awarded a contract in 1974 to build the first eight Block I satellites. In 1978 the contract was extended to build an additional three Block I satellites. Beginning with Navstar 1 in 1978, ten "Block I" GPS satellites were successfully launched. One satellite, "Navstar 7", was lost due to an unsuccessful launch on 18 December 1981.

The Block I satellites were launched from Vandenberg Air Force Base using Atlas rockets that were converted intercontinental ballistic missiles. The satellites were built by Rockwell International at the same plant in Seal Beach, CA where the S-II second stages of the Saturn V rockets had been built.

The Block I series consisted of the concept validation satellites and reflected various stages of system development. Lessons learned from the 11 satellites in the series were incorporated into the fully operational Block II series.

Dual solar arrays supplied over 400 watts of power, charging NiCd batteries for operations in Earth's shadow. S-band communications were used for control and telemetry, while a UHF channel provided cross-links between spacecraft. A hydrazine propulsion system was used for orbital correction. The payload included two L-band navigation signals at 1575.42 MHz (L1) and 1227.60 MHz (L2).

The final Block I launch was conducted on 9 October 1985, but the last Block I satellite was not taken out of service until 18 November 1995, well past its 5-year design life.

## Block II Satellites

### Initial Block II Series

The Block II satellites were the first full scale operational GPS satellites, designed to provide 14 days of operation without any contact from the control segment. The prime contractor was Rockwell International, which built a SVN 12 qualification vehicle after an amendment to the Block I contract. In 1983 the company was awarded an additional contract to build 28 Block II/IIA satellites.

Block II spacecraft were 3-axis stabilized, with ground pointing using reaction wheels. Two solar arrays supplied 710 watts of power, while S band communications were used for control and telemetry. A UHF channel was use for cross-links between spacecraft. A hydrazine propulsion system was used for orbital correction. The payload included two L band navigation signals at 1575.42 MHz (L1) and 1227.60 MHz (L2). Each spacecraft carried 2 rubidium and 2 cesium clocks, as well as nuclear detonation detection sensors, leading to a mass of 1,660 kilograms (3,660 lb).

The first of the nine satellites in the initial Block II series was launched February 14, 1989; the last was launched October 1, 1990. The final satellite of the series to be taken out of service was decommissioned on March 15, 2007, well past its 7.5 year design life.

## Block IIA Series

The Block IIA satellites were slightly improved versions of the Block II series, designed to provide 180 days of operation without contact from the control segment. However, the mass increased to 1,816 kilograms (4,000 lb)

Nineteen satellites in the Block IIA series were launched, the first on November 26, 1990 and the last on November 6, 1997. The final satellite of the Block IIA series was decommissioned on January 25, 2016.

Two of the satellites in this series, numbers 35 and 36, are equipped with laser retro-reflectors, allowing them to be tracked independently of their radio signals, providing unambiguous separation of clock and ephemeris errors.

## Block IIR Series

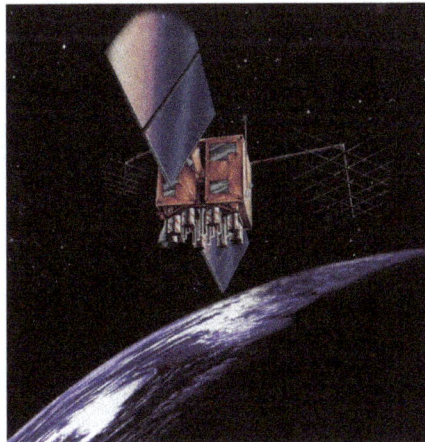

Artist's impression of a GPS-IIR satellite in orbit.

The Block IIR series are "replenishment" satellites developed by Lockheed Martin. Each satellite weighs 4,480 pounds (2,030 kg) at launch and 2,370 pounds (1,080 kg) once on orbit. The first attempted launch of a Block IIR satellite failed on January 17, 1997 when the Delta II rocket exploded 12 seconds into flight. The first successful launch was on July 23, 1997. Twelve satellites in the series were successfully launched.

## Block IIR-M Series

Artist's impression of a GPS-IIRM satellite in orbit.

The Block IIR-M satellites include a new military signal and a more robust civil signal, known as L2C. There are eight satellites in the Block IIR-M series, which were built by Lockheed Martin. The first Block IIR-M satellite was launched on September 26, 2005. The final launch of a IIR-M was on August 17, 2009.

## Block IIF Series

Artist's impression of a Navstar-2F satellite in orbit.

The Block IIF series are "follow-on" satellites developed by Boeing. On September 9, 2007 Boeing announced that it had completed assembly of the first satellite in the Block IIF series. Boeing is under contract to build a total of twelve Block IIF satellites. The first was launched in May 2010 on a Delta IV rocket. The second satellite was launched on July 16, 2011. The satellite has a mass of 1,630 kilograms (3,600 lb) and a design life of 12 years.

## Block IIIA Satellites

The Block IIIA series is the next-generation of GPS satellites, incorporating new signals and broadcasting at higher power levels. In May 2008, Lockheed Martin was awarded a contract for the first two Block III satellites. The contract included options for as many as 10 more. Up to 32 satellites with a design life of 15 years will be procured.

# GPS Signals

Artist's conception of GPS Block II-F satellite in Earth orbit.

Civilian GPS receiver ("GPS navigation device") in a marine application.

Global Positioning System (GPS) satellites broadcast microwave signals to enable GPS receivers on or near the Earth's surface to determine location, velocity, and time. The GPS system itself is operated by the U.S. Department of Defense (DoD) for use by both the military and the general public.

GPS signals include ranging signals, used to measure the distance to the satellite, and navigation messages. The navigation messages include *ephemeris* data, used to calculate the position of each satellite in orbit, and information about the time and status of the entire satellite constellation, called the *almanac*.

There are 4 signals available for civilian use. In order of date of introduction, these are: L1 C/A, L2C, L5 and L1C. L1 C/A is also called the *legacy signal* and is broadcast by all satellites. The other signals are called *modernized signals* and not broadcast by all satellites. In addition, there are *restricted signals*, also broadcast to the general public, but whose encoding is secret and are intended to be used only by authorized parties. Nonetheless, some limited use of restricted signals can be made by civilians without access to the secret encoding details; this is called *codeless* and *semi-codeless* access, and is officially supported.

The interface to the User Segment (GPS receivers) is described in the Interface Control Documents (ICD). The format of civilian signals is described in the Interface Specification (IS) which is a subset of the ICD.

## Common Characteristics

The GPS satellites (called *space vehicles* in the GPS interface specification documents) transmit simultaneously several ranging codes and navigation data using binary phase-shift keying (BPSK). Only a limited number of central frequencies are used; satellites using the same frequency are distinguished by using different ranging codes; in other words, GPS uses code division multiple access. The ranging codes are also called *chipping codes* (in reference to CDMA/DSSS), *pseudorandom noise* and *pseudorandom binary sequences* (in reference to the fact that it is predictable, but statistically it resembles noise).

Some satellites transmit several BPSK streams at the same frequency in quadrature, in a form of quadrature amplitude modulation. However, unlike typical QAM systems where a single bit stream is split in two half-symbol-rate bit streams to improve spectral efficiency, in GPS signals the in-phase and quadrature components are modulated by separate (but functionally related) bit streams.

Satellites are uniquely identified by a serial number called *space vehicle number* (SVN) which does not change during its lifetime. In addition, all operating satellites are numbered with a *space vehicle identifier* (SV ID) and *pseudorandom noise number* (PRN number) which uniquely identifies the ranging codes that a satellite uses. There is a fixed one to one correspondence between SV identifiers and PRN numbers described in the interface specification. Unlike SVNs, the SV ID/PRN number of a satellite may be changed (also changing the ranging codes it uses). At any point in time, any SV ID/PRN number is in use by at most a single satellite. A single SV ID/PRN number may have been used by several satellites at different points in time and a single satellite may have used different SV ID/PRN numbers at different points in time. The current SVNs and PRN numbers for the GPS constellation may be found at NAVCEN.

## Legacy GPS Signals

The original GPS design contains two ranging codes: the *coarse/acquisition* (C/A) code, which is freely available to the public, and the restricted *precision* (P) code, usually reserved for military applications.

## Coarse/Acquisition Code

The C/A PRN codes are Gold codes with a period of 1,023 chips transmitted at 1.023 Mbit/s, implying a period of 1 ms. They are combined with a navigation message using exclusive or and the resulting bit stream is used for modulation as previously described. These codes only match up, or strongly autocorrelate when they are almost exactly aligned. Each satellite uses a unique PRN code, which does not correlate well with any other satellite's PRN code. In other words, the PRN codes are highly orthogonal to one another.

The C/A codes are generated by combining (using "exclusive or") 2 bit streams generated by maximal period 10 stage linear feedback shift registers. Different codes are obtained by selectively delaying one of those bit streams. Thus:

$$C/A_i(t) = A(t) \oplus B(t - D_i)$$

where:

$C/A_i$ is the code with PRN number $i$.

$A$ is the output of the first LFSR whose generator polynomial is $x \rightarrow x^{10} + x^3 + 1$, and initial state is $1111111111_2$.

$B$ is the output of the second LFSR whose generator polynomial is $x \rightarrow x^{10} + x^9 + x^8 + x^6 + x^3 + x^2 + 1$ and initial state is also $1111111111_2$.

$D_i$ is a delay (by an integer number of periods) specific to each PRN number $i$; it is designated in the GPS interface specification.

$\oplus$ is exclusive or.

The arguments of the functions therein are the number of *bits* or *chips* since their epochs, starting at 0. The epoch of the LFSRs is the point at which they are at the initial state; and for the overall C/A codes it the start of any UTC second plus any integer amount of milliseconds. The output of LFSRs at negative arguments is defined consistent with the period which is 1,023 chips (this provision is necessary because $B$ may have a negative argument using the above equation).

The delay for PRN numbers 34 and 37 is the same; therefore their C/A codes are identical and are not transmitted at the same time (it may make one or both of those signals unusable due to mutual interference depending on the relative power levels received on each GPS receiver).

## Precision Code

The P-code is also a PRN; however, each satellite's P-code PRN code is $6.1871 \times 10^{12}$ bits long (6,187,100,000,000 bits, ~720.213 gigabytes) and only repeats once a week (it is transmitted at 10.23 Mbit/s). The extreme length of the P-code increases its correlation gain and eliminates any range ambiguity within the Solar System. However, the code is so long and complex it was believed that a receiver could not directly acquire and synchronize with this signal alone. It was expected that the receiver would first lock onto the relatively simple C/A code and then, after obtaining the current time and approximate position, synchronize with the P-code.

Whereas the C/A PRNs are unique for each satellite, the P-code PRN is actually a small segment of a master P-code approximately $2.35 \times 10^{14}$ bits in length (235,000,000,000,000 bits, ~26.716 terabytes) and each satellite repeatedly transmits its assigned segment of the master code.

To prevent unauthorized users from using or potentially interfering with the military signal through a process called spoofing, it was decided to encrypt the P-code. To that end the P-code was modulated with the *W-code*, a special encryption sequence, to generate the *Y-code*. The Y-code is what the satellites have been transmitting since the anti-spoofing module was set to the "on" state. The encrypted signal is referred to as the *P(Y)-code*.

The details of the W-code are kept secret, but it is known that it is applied to the P-code at approximately 500 kHz, which is a slower rate than that of the P-code itself by a factor of approximately 20. This has allowed companies to develop semi-codeless approaches for tracking the P(Y) signal, without knowledge of the W-code itself.

In addition to the PRN ranging codes, a receiver needs to know detailed information about each satellite's position and the network. The GPS design has this information modulated on top of both the C/A and P(Y) ranging codes at 50 bit/s and calls it the *navigation message*. The navigation message format described in this section is called LNAV data (for *legacy navigation*).

The navigation message conveys information which can be classified in 3 broad areas:

- The GPS date and time, plus the satellite's status and an indication of its health.

- The ephemeris: orbital information which allows the receiver to calculate the position of the satellite. Each satellite transmits its own ephemeris.

- The almanac data: contains information and status concerning all the satellites; each satellite transmits almanac data for several (possibly all) satellites, depending on which PRN numbers are in use.

Whereas ephemeris information is highly detailed and considered valid for no more than four hours, almanac information is more general and is considered valid for up to 180 days. The almanac assists the receiver in determining which satellites to search for, and once the receiver picks up each satellite's signal in turn, it then downloads the ephemeris data directly from that satellite. A position fix using any satellite can not be calculated until the receiver has an accurate and complete copy of that satellite's ephemeris data. If the signal from a satellite is lost while its ephemeris data is being acquired, the receiver must discard that data and start again.

The navigation message consists of 1,500 bit long *frames*. Each frame consists of 5 *subframes* of 300 bits, numbered 1 to 5. In turn each subframe consists of 10 *words* of 30 bit each and requires 6 seconds to transmit. Each subframe has the GPS time.

Subframe 1 contains the GPS date (week number) and information to correct the satellite's time to GPS time, plus satellite status and health. Subframes 2 and 3 together contain the transmitting satellite's ephemeris data. Subframes 4 and 5 contain components of the almanac but each frame contains only 1/25th of the complete almanac; a receiver must process 25 whole frames worth of data to retrieve the entire 15,000 bit almanac message. At this rate, 12.5 minutes are required to receive the entire almanac from a single satellite. Each of the 25 versions of frames 4 and 5 is called a *page* and they are numbered 1 to 25.

Frames begin and end at the start/end of week plus an integer multiple of 30 seconds. At start/end of week the cycling between pages is reset to page 1.

There are 2 navigation message types: LNAV-L is used by satellites with PRN numbers 1 to 32 (called *lower PRN numbers*) and LNAV-U is used by satellites with PRN numbers 33 to 63 (called *upper PRN numbers*). The 2 types use very similar formats. Subframes 1 to 3 are the same while subframes 4 and 5 use almost the same format. Both message types contain almanac data for all satellites using the same navigation message type, but not the other type.

Each subframe begins with a Telemetry Word (TLM), which enables the receiver to detect the beginning of a subframe and determine the receiver clock time at which the navigation subframe begins. The next word is the handover word (HOW), which gives the GPS time (actually the time when the first bit of the next subframe will be transmitted) and identifies the specific subframe within a complete frame. The remaining eight words of the subframe contain the actual data specific to that subframe. Each word includes 6 bits of parity generated using an algorithm based on Hamming codes, which take into account the 24 non-parity bits of that frame and the last 2 bits of the previous word.

After a subframe has been read and interpreted, the time the next subframe was sent can be calculated through the use of the clock correction data and the HOW. The receiver knows the receiver clock time of when the beginning of the next subframe was received from detection of the Telemetry Word thereby enabling computation of the transit time and thus the pseudorange. The receiver is potentially capable of getting a new pseudorange measurement at the beginning of each subframe or every 6 seconds.

## Time

GPS time is expressed with a resolution of 1.5 seconds as a week number and a time of week count (TOW). Its zero point (week 0, TOW 0) is defined to be 1980-01-06T00:00Z. The TOW count is a value ranging from 0 to 403,199 whose meaning is the number of 1.5 second periods elapsed since the beginning of the GPS week. Expressing TOW count thus requires 19 bits ($2^{19}$ = 524,288). GPS time is a continuous time scale in that it does not include leap seconds; therefore the start/end of GPS weeks may differ from that of the corresponding UTC day by an integer amount of seconds.

In each subframe, each hand-over word (HOW) contains the most significant 17 bits of the TOW count corresponding to the start of the next following subframe. Note that the 2 least significant bits can be safely omitted because one HOW occurs in the navigation message every 6 seconds, which is equal to the resolution of the truncated TOW count thereof. Equivalently, the truncated TOW count is the time duration since the last GPS week start/end to the beginning of the next frame in units of 6 seconds.

Each frame contains (in subframe 1) the 10 least significant bits of the corresponding GPS week number. Note that each frame is entirely within one GPS week because GPS frames do not cross GPS week boundaries. Since rollover occurs every 1,024 GPS weeks (approximately every 19.6 years; 1,024 is $2^{10}$), a receiver that computes current calendar dates needs to deduce the upper week number bits or obtain them from a different source. One possible method is for the receiver to save its current date in memory when shut down, and when powered on, assume that the newly decoded truncated week number corresponds to the period of 1,024 weeks that starts at the last saved date. This method correctly deduces the full week number if the receiver is never allowed to remain shut down (or without a time and position fix) for more than 1,024 weeks (~19.6 years).

## Almanac

The *almanac* consists of coarse orbit and status information for each satellite in the constellation, an ionospheric model, and information to relate GPS derived time to Coordinated Universal Time (UTC). Each frame contains a part of the almanac (in subframes 4 and 5) and the complete almanac is transmitted by each satellite in 25 frames total (requiring 12.5 minutes). The almanac serves several purposes. The first is to assist in the acquisition of satellites at power-up by allowing the receiver to generate a list of visible satellites based on stored position and time, while an ephemeris from each satellite is needed to compute position fixes using that satellite. In older hardware, lack of an almanac in a new receiver would cause long delays before providing a valid position, because the search for each satellite was a slow process. Advances in hardware have made the acquisition process much faster, so not having an almanac is no longer an issue. The second purpose is for relating time derived from the GPS (called GPS time) to the international time standard of UTC. Finally, the almanac allows a single-frequency receiver to correct for ionospheric delay error by using a global ionospheric model. The corrections are not as accurate as augmentation systems like WAAS or dual-frequency receivers. However, it is often better than no correction, since ionospheric error is the largest error source for a single-frequency GPS receiver.

## Data Updates

Satellite data is updated typically every 24 hours, with up to 60 days data loaded in case there is a disruption in the ability to make updates regularly. Typically the updates contain new ephemerides, with new almanacs uploaded less frequently. The Control

Segment guarantees that during normal operations a new almanac will be uploaded at least every 6 days.

Satellites broadcast a new ephemeris every two hours. The ephemeris is generally valid for 4 hours, with provisions for updates every 4 hours or longer in non-nominal conditions. The time needed to acquire the ephemeris is becoming a significant element of the delay to first position fix, because as the receiver hardware becomes more capable, the time to lock onto the satellite signals shrinks; however, the ephemeris data requires 18 to 36 seconds before it is received, due to the low data transmission rate.

## Frequency Information

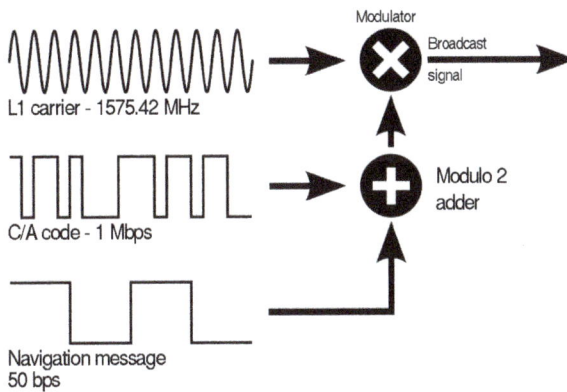

GPS broadcast signal.

For the ranging codes and navigation message to travel from the satellite to the receiver, they must be modulated onto a carrier wave. In the case of the original GPS design, two frequencies are utilized; one at 1575.42 MHz (10.23 MHz × 154) called L1; and a second at 1227.60 MHz (10.23 MHz × 120), called L2.

The C/A code is transmitted on the L1 frequency as a 1.023 MHz signal using a bi-phase shift keying (BPSK) modulation technique. The P(Y)-code is transmitted on both the L1 and L2 frequencies as a 10.23 MHz signal using the same BPSK modulation, however the P(Y)-code carrier is in quadrature with the C/A carrier (meaning it is 90° out of phase).

Besides redundancy and increased resistance to jamming, a critical benefit of having two frequencies transmitted from one satellite is the ability to measure directly, and therefore remove, the ionospheric delay error for that satellite. Without such a measurement, a GPS receiver must use a generic model or receive ionospheric corrections from another source (such as the Wide Area Augmentation System or WAAS). Advances in the technology used on both the GPS satellites and the GPS receivers has made ionospheric delay the largest remaining source of error in the signal. A receiver capable of performing this measurement can be significantly more accurate and is typically referred to as a *dual frequency receiver*.

## Modernization and Additional GPS Signals

Having reached full operational capability on July 17, 1995 the GPS system had completed its original design goals. However, additional advances in technology and new demands on the existing system led to the effort to "modernize" the GPS system. Announcements from the Vice President and the White House in 1998 heralded the beginning of these changes and in 2000, the U.S. Congress reaffirmed the effort, referred to as *GPS III*.

The project involves new ground stations and new satellites, with additional navigation signals for both civilian and military users, and aims to improve the accuracy and availability for all users. A goal of 2013 has been established with incentives offered to the contractors if they can complete it by 2011.

## General Features

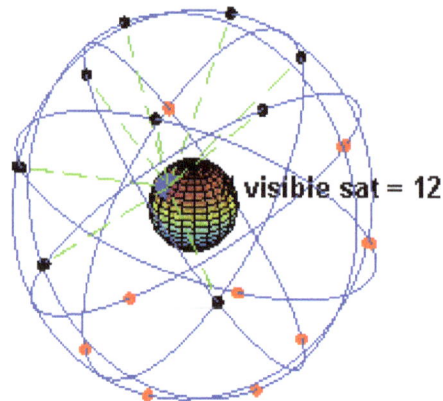

A visual example of the GPS constellation in motion with the Earth rotating. Notice how the number of *satellites in view* from a given point on the Earth's surface, in this example at 45°N, changes with time.

Modernized GPS civilian signals have two general improvements over their legacy counterparts: a dataless acquisition aid and forward error correction (FEC) coding of the NAV message.

A dataless acquisition aid is an additional signal, called a pilot carrier in some cases, broadcast alongside the data signal. This dataless signal is designed to be easier to acquire than the data encoded and, upon successful acquisition, can be used to acquire the data signal. This technique improves acquisition of the GPS signal and boosts power levels at the correlator.

The second advancement is to use forward error correction (FEC) coding on the NAV message itself. Due to the relatively slow transmission rate of NAV data (usually 50 bits per second), small interruptions can have potentially large impacts. Therefore, FEC on the NAV message is a significant improvement in overall signal robustness.

## L2C

One of the first announcements was the addition of a new civilian-use signal, to be transmitted on a frequency other than the L1 frequency used for the coarse/acquisition (C/A) signal. Ultimately, this became the L2C signal, so called because it is broadcast on the L2 frequency. Because it requires new hardware on board the satellite, it is only transmitted by the so-called Block IIR-M and later design satellites. The L2C signal is tasked with improving accuracy of navigation, providing an easy to track signal, and acting as a redundant signal in case of localized interference.

Unlike the C/A code, L2C contains two distinct PRN code sequences to provide ranging information; the *civil-moderate* code (called CM), and the *civil-long* length code (called CL). The CM code is 10,230 bits long, repeating every 20 ms. The CL code is 767,250 bits long, repeating every 1,500 ms. Each signal is transmitted at 511,500 bits per second (bit/s); however, they are multiplexed together to form a 1,023,000 bit/s signal.

CM is modulated with the CNAV Navigation Message, whereas CL does not contain any modulated data and is called a *dataless sequence*. The long, dataless sequence provides for approximately 24 dB greater correlation (~250 times stronger) than L1 C/A-code.

When compared to the C/A signal, L2C has 2.7 dB greater data recovery and 0.7 dB greater carrier-tracking, although its transmission power is 2.3 dB weaker.

## CM and CL Codes

The civil-moderate and civil-long ranging codes are generated by a modular LFSR which is reset periodically to a predetermined initial state. The period of the CM and CL is determined by this resetting and not by the natural period of the LFSR (as is the case with the C/A code). The initial states are designated in the interface specification and are different for different PRN numbers and for CM/CL. The feedback polynomial/mask is the same for CM and CL. The ranging codes are thus given by:

$$\text{CM}_i(t) = A(X_i, t \bmod 10{,}230)$$

$$\text{CL}_i(t) = A(Y_i, t \bmod 767{,}250)$$

where:

$\text{CM}_i$ and $\text{CL}_i$ are the ranging codes for PRN number $i$ and their arguments are the integer number of chips elapsed (starting at 0) since start/end of GPS week, or equivalently since the origin of the GPS time scale.

$X_i$ and $Y_i$ are the initial states for CM and CL respectively. for PRN number $i$.

mod is the remainder of division operation.

$A(x,t)$ is the output of the LFSR when initialized with initial state $x$ after being clocked $t$ times.

$t$ is the integer number of CM and CL chip periods since the origin of GPS time or equivalently, since any GPS second (starting from 0).

The initial states are described in the GPS interface specification as numbers expressed in octal following the convention that the LFSR state is interpreted as the binary representation of a number where the output bit is the least significant bit, and the bit where new bits are shifted in is the most significant bit. Using this convention, the LFSR shifts from most significant bit to least significant bit and when seen in big endian order, it shifts to the right. The states called *final state* in the IS are obtained after 10,229 cycles for CM and after 767,249 cycles for LM (just before reset in both cases).

The feedback bit mask is $100100101001001010100111100_2$. Again with the convention that the least significant bit is the output bit of the LFSR and the most significant bit is the shift-in bit of the LFSR, 0 means no feedback *into* that position, and 1 means feedback *into* that position.

## CNAV Navigation Message

The CNAV data is an upgraded version of the original NAV navigation message. It contains higher precision representation and nominally more accurate data than the NAV data. The same type of information (time, status, ephemeris, and almanac) is still transmitted using the new CNAV format; however, instead of using a frame / subframe architecture, it uses a new pseudo-packetized format made of 12-second 300-bit *messages* analogous to LNAV frames. While LNAV frames have a fixed information content, CNAV messages may be of one of several defined types. The type of a frame determines its information content. Messages do not follow a fixed schedule regarding which message types will be used, allowing the Control Segment some versatility. However, for some message types there are lower bounds on how often they will be transmitted.

In CNAV, at least 1 out of every 4 packets are ephemeris data and the same lower bound applies for clock data packets. The design allows for a wide variety of packet types to be transmitted. With a 32-satellite constellation, and the current requirements of what needs to be sent, less than 75% of the bandwidth is used. Only a small fraction of the available packet types have been defined; this enables the system to grow and incorporate advances without breaking compatibility.

There are many important changes in the new CNAV message:

- It uses forward error correction (FEC) provided by a rate 1/2 convolutional code, so while the navigation message is 25 bit/s, a 50 bit/s signal is transmitted.

- Messages carry a 24 bit CRC, against which integrity can be checked.

- The GPS week number is now represented as 13 bits, or 8192 weeks, and only repeats every 157.0 years, meaning the next return to zero won't occur until the year 2137. This is longer compared to the L1 NAV message's use of a 10-bit week number, which returns to zero every 19.6 years.

- There is a packet that contains a GPS-to-GNSS time offset. This allows better interoperability with other global time-transfer systems, such as Galileo and GLONASS, both of which are supported.

- The extra bandwidth enables the inclusion of a packet for differential correction, to be used in a similar manner to satellite based augmentation systems and which can be used to correct the L1 NAV clock data.

- Every packet contains an alert flag, to be set if the satellite data can not be trusted. This means users will know within 12 seconds if a satellite is no longer usable. Such rapid notification is important for safety-of-life applications, such as aviation.

- Finally, the system is designed to support 63 satellites, compared with 32 in the L1 NAV message.

CNAV messages begin and end at start/end of GPS week plus an integer multiple of 12 seconds. Specifically, the beginning of the first bit (with convolution encoding already applied) to contain information about a message matches the aforesaid synchronization. CNAV messages begin with a 8 bit preamble which is a fixed bit pattern and whose purpose is to enables the receiver to detect the beginning of a message.

## Forward Error Correction Code

The convolutional code used to encode CNAV is described by:

$$X_1(t) = d(t) \oplus d(t-2) \oplus d(t-3) \oplus d(t-5) \oplus d(t-6)$$

$$X_2(t) = d(t) \oplus d(t-1) \oplus d(t-2) \oplus d(t-3) \oplus d(t-6)$$

$$d'(t') = \begin{cases} X_1\left(\dfrac{t'}{2}\right) & \text{if } t' \equiv 0 \pmod 2 \\ X_2\left(\dfrac{t'-1}{2}\right) & \text{if } t' \equiv 1 \pmod 2 \end{cases}$$

where:

$X_1$ and $X_2$ are the unordered outputs of the convolutional encoder

$d$ is the raw (non FEC encoded) navigation data, consisting of the simple concatenation of the 300 bit messages.

*t* is the integer number of *non FEC encoded* navigation data bits elapsed since an arbitrary point in time (starting at 0).

*d'* is the FEC encoded navigation data.

*t'* is the integer number of *FEC encoded* navigation data bits elapsed since the same epoch than *t* (likewise starting at 0).

Since the FEC encoded bit stream runs at 2 times the rate than the non FEC encoded bit as already described, then $t = \left\lfloor \dfrac{t'}{2} \right\rfloor$. FEC encoding is performed independently of navigation message boundaries; this follows from the above equations.

## L2C Frequency Information

An immediate effect of having two civilian frequencies being transmitted is the civilian receivers can now directly measure the ionospheric error in the same way as dual frequency P(Y)-code receivers. However, users utilizing the L2C signal alone, can expect 65% more position uncertainty due to ionospheric error than with the L1 signal alone.

## Military (M-code)

A major component of the modernization process is a new military signal. Called the Military code, or M-code, it was designed to further improve the anti-jamming and secure access of the military GPS signals.

Very little has been published about this new, restricted code. It contains a PRN code of unknown length transmitted at 5.115 MHz. Unlike the P(Y)-code, the M-code is designed to be autonomous, meaning that a user can calculate their position using only the M-code signal. From the P(Y)-code's original design, users had to first lock onto the C/A code and then transfer the lock to the P(Y)-code. Later, direct-acquisition techniques were developed that allowed some users to operate autonomously with the P(Y)-code.

## MNAV Navigation Message

A little more is known about the new navigation message, which is called *MNAV*. Similar to the new CNAV, this new MNAV is packeted instead of framed, allowing for very flexible data payloads. Also like CNAV it can utilize Forward Error Correction (FEC) and advanced error detection (such as a CRC).

## M-code Frequency Information

The M-code is transmitted in the same L1 and L2 frequencies already in use by the previous military code, the P(Y)-code. The new signal is shaped to place most of its energy at the edges (away from the existing P(Y) and C/A carriers).

In a major departure from previous GPS designs, the M-code is intended to be broadcast from a high-gain directional antenna, in addition to a full-Earth antenna. This directional antenna's signal, called a spot beam, is intended to be aimed at a specific region (several hundred kilometers in diameter) and increase the local signal strength by 20 dB, or approximately 100 times stronger. A side effect of having two antennas is that the GPS satellite will appear to be two GPS satellites occupying the same position to those inside the spot beam. While the whole Earth M-code signal is available on the Block IIR-M satellites, the spot beam antennas will not be deployed until the Block III satellites are deployed, tentatively in 2013.

An interesting side effect of having each satellite transmit four separate signals is that the MNAV can potentially transmit four different data channels, offering increased data bandwidth.

The modulation method is binary offset carrier, using a 10.23 MHz subcarrier against the 5.115 MHz code. This signal will have an overall bandwidth of approximately 24 MHz, with significantly separated sideband lobes. The sidebands can be used to improve signal reception.

## L5, Safety of Life

A civilian *safety of life* signal (broadcast in a frequency band protected by the ITU for aeronautical radionavigation service), first broadcast for demonstration purposes on satellite USA-203 (a Block IIR-M series satellite), and available on all GPS IIF satellites (and beyond).

Two PRN ranging codes are transmitted on L5 in quadrature: the in-phase code (called *I5-code*) and the quadrature-phase code (called *Q5-code*). Both codes are 10,230 bits long and transmitted at 10.23 MHz (1 ms repetition period). In addition, I5 is modulated with navigation data (called L5 CNAV) and a 10-bit Neuman-Hoffman code that is clocked at 1 kHz. The Q5-code is additionally modulated with a 20-bit Neuman-Hoffman code that is also clocked at 1 kHz and does not contain navigation data.

Compared to L1 C/A and L2, these are some of the changes in L5:

- Improved signal structure for enhanced performance
- Higher transmitted power than L1/L2 signal (~3 dB, or 2× as powerful)
- Wider bandwidth provides a 10× processing gain, provides sharper autocorrelation (in absolute terms, not relative to chip time duration) and requires a higher sampling rate at the receiver.
- Longer spreading codes (10× longer than C/A)
- Uses the Aeronautical Radionavigation Services band

## I5 and Q5 Codes

The I5-code and Q5-code are generated using the same structure but with different parameters. These codes are the combination (using exclusive or) of the output of 2 LFSR (with different feedback polynomials) which are selectively reset.

$$I5/Q5_i(t) = U(t) \oplus V_i(t)$$

$$U = A((t \bmod 10{,}230) \bmod 8{,}190)$$

$$V_i = B(X_i, t \bmod 10{,}230)$$

where:

$I5/Q5_i$ is the ranging code Q5 or I5 for the PRN number indicated by $i$.

$U$ and $V_i$ are intermediate codes.

$A(x,t')$ is the output of a 13 stage LSFR with feedback polynomial $x^{13} + x^{12} + x^8 + x^7 + x^6 + x^4 + x^3 + x + 1$ and initial state $1111111111111_2$ after being clocked $t'$ times.

$B(x,t')$ is the output of a 13 stage LSFR with feedback polynomial $x^{13} + x^{12} + x^8 + x^7 + x^6 + x^4 + x^3 + x + 1$ and initial state $X_i$ after being clocked $t'$ times.

$X_i$ is the initial state assigned to the PRN number and code $i$ (designated in the IS).

$t$ is the integer number of chip periods since the origin of GPS time or equivalently, since any GPS second (starting from 0).

$i$ is an ordered pair identifying a PRN number and a code (I5 or Q5) and is of the form $(I, n)$ or $(Q, n)$ where $n$ is the PRN number of the satellite, and I, Q are symbols (not variables) that indicate the Q5-code or I5-code, respectively.

$A$ and $B$ are maximal length LFSRs. The modulo operations correspond to resets. Note that both are reset each millisecond (synchronized with C/A code epochs). In addition, the extra modulo operation in the description of $A$ is due to the fact it is reset 1 cycle before its natural period (which is 8,191) so that the next repetition becomes offset by 1 cycle with respect to $B$ (otherwise, since both sequences would repeat, I5 and Q5 would repeat within any 1 ms period as well, degrading correlation characteristics).

## L5 Navigation Message

The L5 CNAV data includes SV ephemerides, system time, SV clock behavior data, status messages and time information, etc. The 50 bit/s data is coded in a rate 1/2 convolution coder. The resulting 100 symbols per second (sps) symbol stream is modulo-2 added to the I5-code only; the resultant bit-train is used to modulate the L5 in-phase (I5) carrier. This combined signal is called the L5 Data signal. The L5 quadrature-phase (Q5) carrier has no data and is called the L5 Pilot signal. The format used for L5 CNAV is very similar to that of L2 CNAV. One difference is that it uses 2 times the data rate. The bit fields within each message, message types, and forward error correction code algorithm are the same as those of L2 CNAV. L5 CNAV messages begin and end at start/end of GPS week plus an integer multiple of 6 seconds (this applies to the beginning of the first bit to contain information about a message, as is the case for L2 CNAV).

## L5 Frequency Information

Broadcast on the L5 frequency (1176.45 MHz, 10.23 MHz × 115), which is an aeronautical navigation band. The frequency was chosen so that the aviation community can manage interference to L5 more effectively than L2.

## L1C

L1C is a civilian-use signal, to be broadcast on the L1 frequency (1575.42 MHz), which contains the C/A signal used by all current GPS users. The L1C will be available with the first Block III launch, tentatively scheduled for the first half of fiscal year 2017.

L1C consists of a pilot (called $L1C_p$) and a data (called $L1C_D$) component. These components use carriers with the same phase (within a margin of error of 100 milliradians), instead of carriers in quadrature as with L5. The PRN codes are 10,230 bits long and transmitted at 1.023 Mbit/s. The pilot component is also modulated by an overlay code called $L1C_O$ (a secondary code that has a lower rate than the ranging code and is also predefined, like the ranging code). Of the total L1C signal power, 25% is allocated to the data and 75% to the pilot. The modulation technique used is BOC(1,1) for the data signal and TMBOC for the pilot. The time multiplexed binary offset carrier (TMBOC) is BOC(1,1) for all except 4 of 33 cycles, when it switches to BOC(6,1).

- Implementation will provide C/A code to ensure backward compatibility

- Assured of 1.5 dB increase in minimum C/A code power to mitigate any noise floor increase

- Data-less signal component pilot carrier improves tracking compared with L1 C/A

- Enables greater civil interoperability with Galileo L1

## L1C Ranging Code

The L1C pilot and data ranging codes are based on a Legendre sequence with length 10,223 used to build an intermediate code (called a *Weil code*) which is expanded with a fixed 7 bit sequence to the required 10,230 bits. This 10,230 bit sequence is the ranging code and varies between PRN numbers and between the pilot and data components. The ranging codes are described by:

$$\text{L1C}_i(t) = \text{L1C}'(t \bmod^* 10{,}230)$$

$$\text{L1C}'_i(t') = \begin{cases} W_i(t') & \text{if } t' < p'_i \\ S(t' - p'_i) & \text{if } p'_i \leq t' < p'_i + 7 \\ W_i(t' - 7) & \text{if } t' \geq p'_i + 7 \end{cases}$$

$$S = (0,1,1,0,1,0,0)$$

$$W_i(n) = L(n) \oplus L((n + w_i) \bmod^* 10{,}223)$$

$$L(n) = \begin{cases} 0 & \text{if } n = 0 \\ 1 & \text{if } n \neq 0 \text{ and there is an integer } m \text{ such that } n \equiv m^2 \ \left( \bmod 10{,}223 \right) \\ 0 & \text{otherwise} \end{cases}$$

where:

is the ranging code for PRN number and component $i$.

$\text{L1C}'_i$ represents a period of $\text{L1C}_i$; it is introduced only to allow a more clear notation. To obtain a direct formula for $\text{L1C}$ start from the right side of the formula for $\text{L1C}'$ and replace all instances of $t'$ with $t \bmod^* 10{,}230$.

$t$ is the integer number of L1C chip periods (which is $(1/1.023)$ µs) since the origin of GPS time or equivalently, since any GPS second (starting from 0).

$i$ is an ordered pair identifying a PRN number and a code ($\text{L1C}_\text{P}$ or $\text{L1C}_\text{D}$) and is of the form $(P, n)$ or $(D, n)$ where $n$ is the PRN number of the satellite, and P, D are symbols (not variables) that indicate the $\text{L1C}_\text{P}$ code or $\text{L1C}_\text{D}$ code, respectively.

$L$ is an intermediate code: a Legendre sequence whose domain is the set of integers $n$ for which $0 \leq n \leq 10{,}222$.

$W_i$ is an intermediate code called Weil code, with the same domain as $L$..

$S$ is a 7 bit long sequence defined for 0-based indexes 0 to 6.

$p'_i$ is the 0-based insertion index of the sequence $S$ into the ranging code (specific for PRN number and code $i$). It is defined in the Interface Specification (IS) as a 1-based index $p$, therefore $p'_i = p_i - 1$.

$w_i$ is the Weil index for PRN number and code $i$ designated in the IS.

mod* is the remainder of integer division (or modulo) operation (not to be confused with the notation for modular congruence, also used in these formulas).

According to the formula above and the GPS IS, the first $w_i$ bits (equivalently, up to the insertion point of $S$ ) of $L1C'_i$ and $L1C$ are the first bits the corresponding Weil code; the next 7 bits are $S$; the remaining bits are the remaining bits of the Weil code.

The IS asserts that $0 \leq p'_i \leq 10{,}222$. For clarity, the formula for $L1C'_i$ does not account for the hypothetical case in which $p'_i > 10{,}222$, which would cause the instance of $S$ inserted into $L1C'_i$ to wrap from index 10,229 to 0.

## L1C Overlay Code

The overlay codes are 1,800 bits long and is transmitted at 100 bit/s, synchronized with the navigation message encoded in $L1C_D$.

For PRN numbers 1 to 63 they are the truncated outputs of maximal period LFSRs which vary in initial conditions and feedback polynomials.

For PRN numbers 64 to 210 they are truncated Gold codes generated by combining 2 LFSR outputs ($S1_i$ and $S2_i$, where $i$ is the PRN number) whose initial state varies. $S1_i$ has one of the 4 feedback polynomials used overall (among PRN numbers 64–210). $S2_i$ has the same feedback polynomial for all PRN numbers in the range 64–210.

## CNAV-2 Navigation Message

| Subframes | | | |
|---|---|---|---|
| **Subframe** | **Bit count** | | **Description** |
| | **Raw** | **Encoded** | |
| 1 | 9 | 52 | Time of interval (TOI) |
| 2 | 576 | 1,200 | Time correction and ephemeris data |
| 3 | 250 | 548 | Variable data |

| Subframe 3 pages | |
|---|---|
| **Page no.** | **Description** |
| 1 | UTC & IONO |
| 2 | GGTO & EOP |
| 3 | Reduced almanac |
| 4 | Midi almanac |
| 5 | Differential correction |
| 6 | Text |

The L1C navigation data (called CNAV-2) is broadcast in 1,800 bits long (including FEC) frames and is transmitted at 100 bit/s.

The frames of L1C are analogous to the messages of L1C and L5. While L1 CNAV and L5 CNAV use a dedicated message type for ephemeris data, all CNAV-2 frames include that information.

The common structure of all messages consists of 3 frames, as listed in the table to the right. The content of subframe 3 varies according to its page number which is analogous to the type number of L1 CNAV and L5 CNAV messages. Pages are broadcast in an arbitrary order.

The time of messages is expressed in a different format than the format of the previous civilian signals. Instead it consists of 3 components:

1. The week number, with the same meaning as with the other civilian signals. Each message contains the week number modulo 8,192 or equivalently, the 13 least significant bits of the week number.

2. An interval time of week (ITOW): the integer number of 2 hour periods elapsed since the latest start/end of week. It has range 0 to 83 (inclusive).

3. A time of interval (TOI): the integer number of 18 second periods elapsed since the period represented by the current ITOW to the beginning of the *next* message. It has range 0 to 399 (inclusive).

TOI is the only content of subframe 1. The week number and ITOW are contained in subframe 2 along with other information.

Subframe 1 is encoded by a modified BCH code. Specifically, the 8 least significant bits are BCH encoded to generate 51 bits, then combined using exclusive or with the most significant bit and finally the most significant bit is appended as the most significant bit of the previous result to obtain the final 52 bits. Subframes 2 and 3 are individually expanded with a 24 bit CRC, then individually encoded using a low-density parity-check code, and then interleaved as a single unit using a block interleaver.

## Frequencies Used by GPS

| GPS Frequencies | | | | |
|---|---|---|---|---|
| **Band** | **Frequency (MHz)** | **Phase** | **Original usage** | **Modernized usage** |
| L1 | 1575.42 (10.23×154) | In-phase (I) | Encrypted precision P(Y) code | |
| | | Quadrature-phase (Q) | Coarse/acquisition (C/A) code | C/A, L1 Civilian (L1C), and Military (M) code |

| L2 | 1227.60 (10.23×120) | In-phase (I) | Encrypted precision P(Y) code | |
| | | Quadrature-phase (Q) | Unmodulated carrier | L2 Civilian (L2C) code and Military (M) code |
| L3 | 1381.05 (10.23×135) | | Used by Nuclear Detonation (NUDET) Detection System Payload (NDS); signals nuclear detonations/ high-energy infrared events. Used to enforce nuclear test ban treaties. | |
| L4 | 1379.913 (10.23×1214/9) | | *(No transmission)* | Being studied for additional ionospheric correction |
| L5 | 1176.45 (10.23×115) | In-phase (I) | *(No transmission)* | Safety-of-Life (SoL) Data signal |
| | | Quadrature-phase (Q) | | Safety-of-Life (SoL) Pilot signal |

All satellites broadcast at the same two frequencies, 1.57542 GHz (L1 signal) and 1.2276 GHz (L2 signal). The satellite network uses a CDMA spread-spectrum technique where the low-bitrate message data is encoded with a high-rate pseudo-random (PRN) sequence that is different for each satellite. The receiver must be aware of the PRN codes for each satellite to reconstruct the actual message data. The C/A code, for civilian use, transmits data at 1.023 million chips per second, whereas the P code, for U.S. military use, transmits at 10.23 million chips per second. The L1 carrier is modulated by both the C/A and P codes, while the L2 carrier is only modulated by the P code. The P code can be encrypted as a so-called P(Y) code which is only available to military equipment with a proper decryption key. Both the C/A and P(Y) codes impart the precise time-of-day to the user.

Each composite signal (in-phase and quadrature phase) becomes:

$$S(t) = \sqrt{P_I}\, X_I(t) \cos(\omega t + \phi_0) \underbrace{- \sqrt{P_Q}\, X_Q(t) \sin(\omega t + \phi_0)}_{+\sqrt{P_Q}\, X_Q(t) \cos\left(\omega t + \phi_0 + \frac{\pi}{2}\right)},$$

where $P_I$ and $P_Q$ represent signal powers; $X_I(t)$ and $X_Q(t)$ represent codes with/without data $(=\pm 1)$. This is a formula for the ideal case (which is not attained in practice) as it does not model timing errors, noise, amplitude mismatch between components or quadrature error (when components are not exactly in quadrature).

## Demodulation and Decoding

A GPS receiver processes the GPS signals received on its antenna to determine position, velocity and/or timing. The signal at antenna is amplified, down converted to baseband or intermediate frequency, filtered (to remove frequencies outside the intended frequency range for the digital signal that would alias into it) and digitalized; these steps may be chained in a different order. Note that aliasing is sometimes intentional (specifically, when undersampling is used) but filtering is still required to discard frequencies not intended to be present in the digital representation.

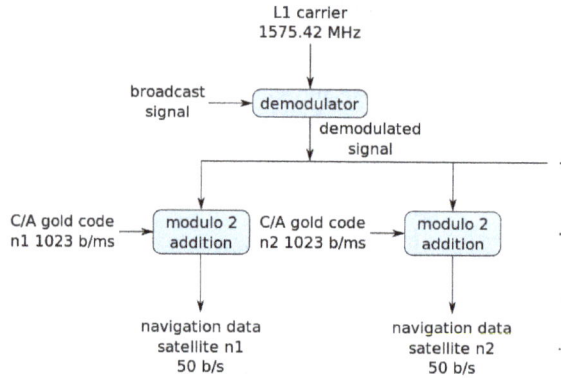

Demodulating and Decoding GPS Satellite Signals using the Coarse/Acquisition Gold code.

For each satellite used by the receiver, the receiver must first acquire the signal and then track it as long as that satellite is in use; both are performed in the digital domain in by far most (if not all) receivers.

Acquiring a signal is the process of determining the frequency and code phase (both relative to receiver time) when it was previously unknown. Code phase must be determined within an accuracy that depends on the receiver design (especially the tracking loop); 0.5 times the duration of code chips (approx. 0.489 μs) is a representative value.

Tracking is the process of continuously adjusting the estimated frequency and phase to match the received signal as close as possible and therefore is a phase locked loop. Note that acquisition is performed to start using a particular satellite, but tracking is performed as long as that satellite is in use.

In this section, one possible procedure is described for L1 C/A acquisition and tracking, but the process is very similar for the other signals. The described procedure is based on computing the correlation of the received signal with a locally generated replica of the ranging code and detecting the highest peak or lowest valley. The offset of the highest peak or lowest valley contains information about the code phase relative to receiver time. The duration of the local replica is set by receiver design and is typically shorter than the duration of navigation data bits, which is 20 ms.

## Acquisition

Acquisition of a given PRN number can be conceptualized as searching for a signal in a bidimensional search space where the dimensions are (1) code phase, (2) frequency. In addition, a receiver may not know which PRN number to search for, and in that case a third dimension is added to the search space: (3) PRN number.

Frequency space: The frequency range of the search space is the band where the signal may be located given the receiver knowledge. The carrier frequency varies by roughly 5 kHz due to the Doppler effect when the receiver is stationary; if the receiver moves, the variation is higher. The code frequency deviation is 1/1,540 times the carrier frequency deviation for L1 because the code frequency is 1/1,540 of the carrier frequency. The down conversion does not affect the frequency deviation; it only shifts all the signal frequency components down. Since the frequency is referenced to the receiver time, the uncertainty in the receiver oscillator frequency adds to the frequency range of the search space.

Code phase space: The ranging code has a period of 1,023 chips each of which lasts roughly 0.977 µs. The code gives strong autocorrelation only at offsets less than 1 in magnitude. The extent of the search space in the code phase dimension depends on the granularity of the offsets at which correlation is computed. It is typical to search for the code phase within a granularity of 0.5 chips or finer; that means 2,046 offsets. There may be more factors increasing the size of the search space of code phase. For example, a receiver may be designed so as to examine 2 consecutive windows of the digitalized signal, so that at least one of them does not contain a navigation bit transition (which worsens the correlation peak); this requires the signal windows to be at most 10 ms long.

PRN number space: The lower PRN numbers range from 1 to 32 and therefore there are 32 PRN numbers to search for when the receiver does not have information to narrow the search in this dimension. The higher PRN numbers range from 33 to 66.

If the almanac information has previously been acquired, the receiver picks which satellites to listen for by their PRNs. If the almanac information is not in memory, the receiver enters a search mode and cycles through the PRN numbers until a lock is obtained on one of the satellites. To obtain a lock, it is necessary that there be an unobstructed line of sight from the receiver to the satellite. The receiver can then decode the almanac and determine the satellites it should listen for. As it detects each satellite's signal, it identifies it by its distinct C/A code pattern.

## Simple Correlation

The simplest way to acquire the signal (not necessarily the most effective or least computationally expensive) is to compute the dot product of a window of the digitalized signal with a set of locally generated replicas. The locally generated replicas vary in carrier frequency and code phase to cover all the already mentioned search space which is the Cartesian product of the frequency search space and the code phase search space. The carrier is a complex number where real and imaginary components are both sinusoids as described by Euler's formula. The replica that generates the highest magnitude of dot product is likely the one that best matches the code phase and frequency of the signal; therefore, if that magnitude is above a threshold, the receiver proceeds to track the signal or further refine the estimated parameters before tracking. The threshold is used to minimize false positives (apparently detecting a signal when there is in fact no signal), but some may still occur occasionally.

Using a complex carrier allows the replicas to match the digitalized signal regardless of the signal's carrier phase and to detect that phase (the principle is the same used by the Fourier transform). The dot product is a complex number; its magnitude represents the level of similarity between the replica and the signal, as with an ordinary correlation of real-valued time series. The argument of the dot product is an approximation of the corresponding carrier in the digitalized signal.

As an example, assume that the granularity for the search in code phase is 0.5 chips and in frequency is 500 Hz, then there are 1,023/0.5=2,046 code phases and 10,000 Hz/500 Hz=20 frequencies to try for a total of 20×2,046=40,920 local replicas. Note that each frequency bin is centered on its interval and therefore covers 250 Hz in each direction; for example, the first bin has a carrier at −4,750 Hz and covers the interval −5,000 Hz to −4,500 Hz. Code phases are equivalent modulo 1,023 because the ranging code is perodic; for example, phase −0.5 is equivalent to phase 1,022.5.

The following table depicts the local replicas that would be compared against the digitalized signal in this example. "•" means a single local replica while "..." is used for elided local replicas:

| Carrier freq. deviation | Code phase (in chips) | | | | |
|---|---|---|---|---|---|
| | 0.0 | 0.5 | (more phases) | 1,022.0 | 1,022.5 |
| −4,750 Hz | • | • | ... | • | • |
| −4,250 Hz | • | • | ... | • | • |
| (more frequencies) | ... | ... | ... | ... | ... |
| 4,250 Hz | • | • | ... | • | • |
| 4,750 Hz | • | • | ... | • | • |

## Fourier Transform

As an improvement over the simple correlation method, it is possible to implement the computation of dot products more efficiently with a Fourier transform. Instead of performing one dot product for each element in the Cartesian product of code and frequency, a single operation involving FFT and covering all frequencies is performed for each code phase; each such operation is more computationally expensive, but it may still be faster overall than the previous method due to the efficiency of FFT algorithms, and it recovers carrier frequency with a higher accuracy, because the frequency bins are much closely spaced in a DFT.

Specifically, for all code phases in the search space, the digitalized signal window is multiplied element by element with a local replica of the code (with no carrier), then processed with a discrete Fourier transform.

Given the previous example to be processed with this method, assume real-valued data (as opposed to complex data, which would have in-phase and quadrature components), a sampling rate of 5 MHz, a signal window of 10 ms, and an intermediate frequency of 2.5 MHz. There will be 5 MHz×10 ms=50,000 samples in the digital signal, and therefore 25,001 frequency components ranging from 0 Hz to 2.5 MHz in steps of 100 Hz (note that the 0 Hz component is real because it is the average of a real-valued signal and the 2.5 MHz component is real as well because it is the critical frequency). Only the components (or bins) within 5 kHz of the central frequency are examined, which is the range from 2.495 MHz to 2.505 MHz, and it is covered by 51 frequency components. There are 2,046 code phases as in the previous case, thus in total 51×2,046=104,346 complex frequency components will be examined.

## Circular Correlation with Fourier Transform

Likewise, as an improvement over the simple correlation method, it is possible to perform a single operation covering all code phases for each frequency bin. The operation performed for each code phase bin involves forward FFT, element-wise multiplication in the frequency domain. inverse FFT, and extra processing so that overall, it computes circular correlation instead of circular convolution. This yields more accurate *code phase determination* than the simple correlation method in contrast with the previous method, which yields more accurate *carrier frequency determination* than the previous method.

## Tracking and Navigation Message Decoding

Since the carrier frequency received can vary due to Doppler shift, the points where received PRN sequences begin may not differ from O by an exact integral number of milliseconds. Because of this, carrier frequency tracking along with PRN code tracking are used to determine when the received satellite's PRN code begins. Unlike the earlier computation of offset in which trials of all 1,023 offsets could potentially be required, the tracking to maintain lock usually requires shifting of half a pulse width or less. To perform this tracking, the receiver observes two quantities, phase error and received frequency offset. The correlation of the received PRN code with respect to the receiver generated PRN code is computed to determine if the bits of the two signals are misaligned. Comparisons of the received PRN code with receiver generated PRN code shifted half a pulse width early and half a pulse width late are used to estimate adjustment required. The amount of adjustment required for maximum correlation is used in estimating phase error. Received frequency offset from the frequency generated by the receiver provides an estimate of phase rate error. The command for the frequency generator and any further PRN code shifting required are computed as a function of the phase error and the phase rate error in accordance with the control law used. The Doppler velocity is computed as a function of the frequency offset from the carrier nominal frequency. The Doppler velocity is the velocity component along the line of sight of the receiver relative to the satellite.

As the receiver continues to read successive PRN sequences, it will encounter a sudden change in the phase of the 1,023 bit received PRN signal. This indicates the beginning of a data bit of the navigation message. This enables the receiver to begin reading the 20 millisecond bits of the navigation message. The TLM word at the beginning of each subframe of a navigation frame enables the receiver to detect the beginning of a subframe and determine the receiver clock time at which the navigation subframe begins. The HOW word then enables the receiver to determine which specific subframe is being transmitted. There can be a delay of up to 30 seconds before the first estimate of position because of the need to read the ephemeris data before computing the intersections of sphere surfaces.

After a subframe has been read and interpreted, the time the next subframe was sent can be calculated through the use of the clock correction data and the HOW. The receiver knows the receiver clock time of when the beginning of the next subframe was received from detection of the Telemetry Word thereby enabling computation of the transit time and thus the pseudorange. The receiver is potentially capable of getting a new pseudorange measurement at the beginning of each subframe or every 6 seconds.

Then the orbital position data, or ephemeris, from the navigation message is used to calculate precisely where the satellite was at the start of the message. A more sensitive receiver will potentially acquire the ephemeris data more quickly than a less sensitive receiver, especially in a noisy environment.

### Sources and References

### Bibliography

### GPS Interface Specification

- *"GPS Interface Specification (GPS-IS-200H)" (PDF). 24 September 2013.* (describes L1, L2C and P).

- *"GPS Interface Specification (GPS-IS-705D)" (PDF). 24 September 2013.* (describes L5).

- *"GPS Interface Specification (GPS-IS-800D)" (PDF). 24 September 2013.* (describes L1C).

# GNSS Positioning Calculation

The global navigation satellite system (GNSS) positioning for receiver's position is derived through the calculation steps, or algorithm, given below. In essence, a GNSS receiver measures the transmitting time of GNSS signals emitted from four or more GNSS satellites and these measurements are used to obtain its position (i.e., spatial coordinates) and reception time.

## Calculation Steps

1. A global-navigation-satellite-system (GNSS) receiver measures the apparent transmitting time, $\tilde{t}_i$, or "phase", of GNSS signals emitted from four or more GNSS satellites ($i = 1,2,3,4,..,n$), simultaneously.

2. GNSS satellites broadcast the messages of satellites' ephemeris, $r_i(t)$, and intrinsic clock bias (i.e., clock advance), $\delta t_{clock,sv,i}(t)$ as the functions of (atomic) standard time, e.g., GPST.

3. The transmitting time of GNSS satellite signals, $t_i$, is thus derived from the non-closed-form equations $\tilde{t}_i = t_i + \delta t_{clock,i}(t_i)$ and $\delta t_{clock,i}(t_i) = \delta t_{clock,sv,i}(t_i) + \delta t_{orbit\text{-}relativ,i}(\mathbf{r}_i,\mathbf{r}_i),$, where $\delta t_{orbit\text{-}relativ,i}(\mathbf{r}_i,\mathbf{r}_i)$ is the relativistic clock bias, periodically risen from the satellite's orbital eccentricity and Earth's gravity field. The satellite's position and velocity are determined by $t_i$ as follows: $r_i = r_i(t_i)$ and $\mathbf{r}_i = \mathbf{r}_i(t_i)$.

4. In the field of GNSS, "geometric range", $r(\mathbf{r}_A, \mathbf{r}_B)$, is defined as straight range, or 3-dimensional distance, from $\mathbf{r}_A$ to $\mathbf{r}_B$ in inertial frame (e.g., Earth-centered inertial (ECI) one), not in rotating frame.

5. The receiver's position, $\mathbf{r}_{rec}$, and reception time, $t_{rec}$, satisfy the light-cone equation of $r(\mathbf{r}_i, \mathbf{r}_{rec})/c + (t_i - t_{rec}) = 0$ in inertial frame, where $c$ is the speed of light. The signal transit time is $-(t_i - t_{rec})$.

6. The above is extended to the satellite-navigation positioning equation, $r(\mathbf{r}_i, \mathbf{r}_{rec})/c + (t_i - t_{rec}) + \delta t_{atmos,i} - \delta t_{meas\text{-}err,i} = 0$, where $\delta t_{atmos,i}$ is atmospheric delay (= ionospheric delay + tropospheric delay) along signal path and $\delta t_{meas\text{-}err,i}$ is the measurement error.

7. The Gauss–Newton method can be used to solve the nonlinear least-squares problem for the solution: $(\hat{\mathbf{r}}_{rec}, \hat{t}_{rec}) = \arg\min \phi(\mathbf{r}_{rec}, t_{rec})$, where $\phi(\mathbf{r}_{rec}, t_{rec}) = \sum_{i=1}(\delta t_{meas\text{-}err,i}/\sigma_{\delta t_{meas\text{-}err,i}})^2$. Note that $\delta t_{meas\text{-}err,i}$ should be regarded as a function of $\mathbf{r}_{rec}$ and $t_{rec}$.

8. The posterior distribution of $r_{rec}$ and $t_{rec}$ is proportional to $\exp(-\frac{1}{2}\phi(\mathbf{r}_{rec}, t_{rec}))$, whose mode is $(\hat{\mathbf{r}}_{rec}, \hat{t}_{rec})$.. Their inference is formalized as maximum a posteriori estimation.

9. The posterior distribution of $\mathbf{r}_{rec}$ is proportional to $\begin{cases} \Delta t_i(t_i, E_i) \triangleq t_i + \delta t_{clock,i}(t_i, E_i) - \tilde{t}_i = 0, \\ \Delta M_i(t_i, E_i) \triangleq M_i(t_i) - (E_i - e_i \sin E_i) = 0, \end{cases}$ .

## The Solution Illustrated

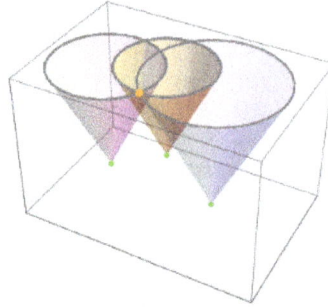

Essentially, the solution, , is the intersection of light cones.

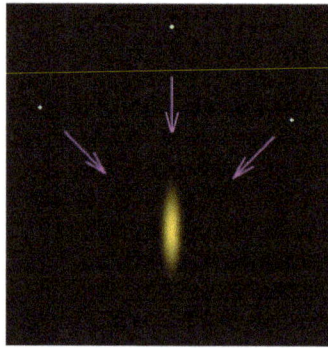

The posterior distribution of the solution is derived from
the product of the distribution of propagating spherical surfaces.

## The GPS Case

- For Global Positioning System (GPS), the non-closed-form equations in step 3 result in

$$\begin{cases} \Delta t_i(t_i, E_i) \triangleq t_i + \delta t_{\text{clock},i}(t_i, E_i) - \tilde{t}_i = 0, \\ \Delta M_i(t_i, E_i) \triangleq M_i(t_i) - (E_i - e_i \sin E_i) = 0, \end{cases}$$

in which $E_i$ is the orbital eccentric anomaly of satellite $i$, $M_i$ is the mean anomaly, $e_i$ is the eccentricity, and $\delta t_{\text{clock},i}(t_i, E_i) = \delta t_{\text{clock,sv},i}(t_i) + \delta t_{\text{orbit-relativ},i}(E_i)$.

- The above can be solved by using the bivariate Newton–Raphson method on $t_i$ and $E_i$. Two times of iteration will be necessary and sufficient in most cases. Its iterative update will be described by using the approximated inverse of Jacobian matrix as follows:

$$\begin{pmatrix} t_i \\ E_i \end{pmatrix} \leftarrow \begin{pmatrix} t_i \\ E_i \end{pmatrix} - \begin{pmatrix} 1 & 0 \\ \dfrac{\dot{M}_i(t_i)}{1 - e_i \cos E_i} & -\dfrac{1}{1 - e_i \cos E_i} \end{pmatrix} \begin{pmatrix} \Delta t_i \\ \Delta M_i \end{pmatrix}$$

- Tropospheric delay should not be ignored, while the Global Positioning System (GPS) specification doesn't provide its detailed description.

## The GLONASS Case

- The GLONASS ephemerides don't provide clock biases $\delta t_{\text{clock,sv},i}(t)$, but $\delta t_{\text{clock},i}(t)$.

# GNSS Enhancement

GNSS enhancement refers to techniques used to improve the accuracy of positioning information provided by the Global Positioning System or other global navigation satellite systems in general, a network of satellites used for navigation.

Enhancement methods of improving accuracy rely on external information being integrated into the calculation process. There are many such systems in place and they are generally named or described based on how the GPS sensor receives the information. Some systems transmit additional information about sources of error (such as clock drift, ephemeris, or ionospheric delay), others provide direct measurements of how much the signal was off in the past, while a third group provide additional navigational or vehicle information to be integrated in the calculation process.

Examples of augmentation systems include the Wide Area Augmentation System, Differential GPS, Inertial Navigation Systems and Assisted GPS.

## Background

The Global Positioning System (GPS) is a satellite-based system for navigation. Receivers on or near the earth's surface can determine their locations based on signals received from any four or more of the satellites in the network.

All satellites in the world broadcast on the same two frequencies, known as L1 (1575.42 MHz) and L2 (1227.60 MHz). The network uses code division multiple access (CDMA) to allow separate messages from the individual satellites to be distinguished. Two distinct CDMA encodings are used: the coarse/acquisition (C/A) code, which is accessible by the general public, and the precise (P) code, that is encrypted so that only the U.S. military can access it. The messages sent from each satellite contain information ranging from the satellite health, the satellite's orbital path, the clock state of the satellite, and the configuration of the entire satellite network.

## Precise Monitoring

The accuracy of a calculation can also be improved through precise monitoring and measuring of the existing GPS signals in additional or alternate ways.

After *Selective Availability* was turned off by the U.S. government, the largest error in GPS was usually the unpredictable delay through the ionosphere. The spacecraft broadcast ionospheric model parameters, but they are necessarily imperfect. This is one reason the GPS spacecraft transmit on at least two frequencies, L1 and L2. Ionospheric delay is a well-defined function of frequency and the total electron content (TEC) along the path, so measuring the arrival time difference between the frequencies determines TEC and thus the precise ionospheric delay at each frequency.

Receivers with decryption keys can decode the P(Y)-code transmitted on both L1 and L2. However, these keys are reserved for the military and authorized agencies and are not available to the public. Without keys, it is still possible to use a *codeless* technique to compare the P(Y) codes on L1 and L2 to gain much of the same error information. However, this technique is slow, so it is currently limited to specialized surveying equipment. In the future, additional civilian codes are expected to be transmitted on the L2 and L5 frequencies. Then all users will be able to perform dual-frequency measurements and directly compute ionospheric delay errors.

A second form of precise monitoring is called *Carrier-Phase Enhancement* (CPGPS). The error, which this corrects, arises because the pulse transition of the PRN is not instantaneous, and thus the correlation (satellite-receiver sequence matching) operation is imperfect. The CPGPS approach utilizes the L1 carrier wave, which has a period of

$$\frac{1\,s}{1575.42\times10^6} = 0.63475\,ns \approx 1\,ns$$

which is about one-thousandth of the C/A Gold code bit period of

$$\frac{1\,s}{1023\times10^3} = 977.5\,ns \approx 1000\,ns$$

to act as an additional clock signal and resolve the uncertainty. The phase difference error in the normal GPS amounts to between 2 and 3 meters (6 to 10 ft) of ambiguity. CPGPS working to within 1% of perfect transition reduces this error to 3 centimeters (1 inch) of ambiguity. By eliminating this source of error, CPGPS coupled with DGPS normally realizes between 20 and 30 centimeters (8 to 12 inches) of absolute accuracy.

## Real-time Kinematic Positioning

*Real-time Kinematic Positioning* (RKP) is another approach for a precise GPS-based positioning system. In this approach, determination of range signal can be resolved to a precision of less than 10 centimeters (4 in). This is done by resolving the number of cycles in which the signal is transmitted and received by the receiver. This can be accomplished by using a combination of differential GPS (DGPS) correction data, transmitting GPS signal phase information and ambiguity resolution techniques via statistical tests—possibly with processing in real-time.

# Timekeeping

While most clocks are synchronized to Coordinated Universal Time (UTC), the atomic clocks on the satellites are set to *GPS time*. The difference is that GPS time is not corrected to match the rotation of the Earth, so it does not contain leap seconds or other corrections which are periodically added to UTC. GPS time was set to match Coordinated Universal Time (UTC) in 1980, but has since diverged. The lack of corrections means that GPS time remains at a constant offset with International Atomic Time (TAI) (TAI - GPS = 19 seconds). Periodic corrections are performed on the on-board clocks to correct relativistic effects and keep them synchronized with ground clocks.

The GPS navigation message includes the difference between GPS time and UTC, which as of 2016 is 17 seconds due to leap seconds added since 1980. Receivers subtract this offset from GPS time to calculate UTC and specific timezone values. The GPS-UTC offset, a signed eight-bit field, can accommodate up to 128 negative leaps or 127 positive leap seconds. At the current rate of Earth's rotation (with one positive leap second introduced approximately every 18 months), the offset field is sufficient for approximately one more century. However, as the Earth is not only rotating slower than when UTC was adopted in 1972, but also slowing down at a somewhat unpredictable rate, leap seconds will begin to occur more frequently. So, the GPS offset field may overflow within the 21st century.

As opposed to the year, month, and day format of the Gregorian calendar, the GPS date is expressed as a week number and a seconds-into-week number. The week number is transmitted as a ten-bit field in the C/A and P(Y) navigation messages, and so it becomes zero again every 1,024 weeks (19.6 years). GPS week zero started at 00:00:00 UTC (00:00:19 TAI) on January 6, 1980, and the week number became zero again for the first time at 23:59:47 UTC on August 21, 1999 (00:00:19 TAI on August 22, 1999). To determine the current Gregorian date, a GPS receiver must be provided with the approximate date (to within 3,584 days) to correctly translate the GPS date signal. To address this concern the modernized GPS navigation message uses a 13-bit field, which only repeats every 8,192 weeks (157 years), thus lasting until the year 2137 (157 years after GPS week zero).

# Carrier Phase Tracking (Surveying)

Utilizing the navigation message to measure pseudorange has been discussed. Another method that is used in GPS surveying applications is carrier phase tracking. The period of the carrier frequency times the speed of light gives the wavelength, which is about 0.19 meters for the L1 carrier. With a 1% of wavelength accuracy in detecting the leading edge, this component of pseudorange error might be as low as 2 millimeters. This compares to 3 meters for the C/A code and 0.3 meters for the P code.

However, this 2 millimeter accuracy requires measuring the total phase, that is the total number of wavelengths plus the fractional wavelength. This requires specially equipped receivers. This method has many applications in the field of surveying.

We now describe a method which could potentially be used to estimate the position of receiver 2 given the position of receiver 1 using triple differencing followed by numerical root finding, and a mathematical technique called least squares. A detailed discussion of the errors is omitted in order to avoid detracting from the description of the methodology. In this description differences are taken in the order of differencing between satellites, differencing between receivers, and differencing between epochs. This should not be construed to mean that this is the only order which can be used. Indeed, other orders of taking differences are equally valid.

The satellite carrier total phase can be measured with ambiguity as to the number of cycles. Let $\phi(r_i, s_j, t_k)$ denote the phase of the carrier of satellite j measured by receiver i at time $t_k$. This notation has been chosen so as to make it clear what the subscripts i, j, and k mean. In view of the fact that the receiver, satellite, and time come in alphabetical order as arguments of $\phi$ and to strike a balance between readability and conciseness, let $\phi_{i,j,k} = \phi(r_i, s_j, t_k)$ so as to have a concise abbreviation. Also we define three functions, : $\Delta^r, \Delta^s, \Delta^t$ which perform differences between receivers, satellites, and time points respectively. Each of these functions has a linear combination of variables with three subscripts as its argument. These three functions are defined below. If $\alpha_{i,j,k}$ is a function of the three integer arguments, i, j, and k then it is a valid argument for the functions, : $\ddot{A}^r, \ddot{A}^s, \ddot{A}^t$, with the values defined as

$$\Delta^r(\alpha_{i,j,k}) = \alpha_{i+1,j,k} - \alpha_{i,j,k},$$

$$\Delta^s(\alpha_{i,j,k}) = \alpha_{i,j+1,k} - \alpha_{i,j,k}, \text{ and}$$

$$\Delta^t(\alpha_{i,j,k}) = \alpha_{i,j,k+1} - \alpha_{i,j,k}.$$

Also if $\alpha_{i,j,k}$ and $\beta_{l,m,n}$ are valid arguments for the three functions and a and b are constants then $(a\,\alpha_{i,j,k} + b\,\beta_{l,m,n})$ is a valid argument with values defined as

$$\Delta^r(a\,\alpha_{i,j,k} + b\,\beta_{l,m,n}) = a\,\Delta^r(\alpha_{i,j,k}) + b\,\Delta^r(\beta_{l,m,n}),$$

$$\Delta^s(a\,\alpha_{i,j,k} + b\,\beta_{l,m,n}) = a\,\Delta^s(\alpha_{i,j,k}) + b\,\Delta^s(\beta_{l,m,n}), \text{ and}$$

$$\Delta^t(a\,\alpha_{i,j,k} + b\,\beta_{l,m,n}) = a\,\Delta^t(\alpha_{i,j,k}) + b\,\Delta^t(\beta_{l,m,n}).$$

Receiver clock errors can be approximately eliminated by differencing the phases measured from satellite 1 with that from satellite 2 at the same epoch. This difference is designated as $\Delta^s(\phi_{1,1,1}) = \phi_{1,2,1} - \phi_{1,1,1}$

Double differencing can be performed by taking the differences of the between satellite difference observed by receiver 1 with that observed by receiver 2. The satellite clock errors will be approximately eliminated by this between receiver differencing. This double difference is:

$$\Delta^r(\Delta^s(\phi_{1,1,1})) = \Delta^r(\phi_{1,2,1} - \phi_{1,1,1}) = \Delta^r(\phi_{1,2,1}) - \Delta^r(\phi_{1,1,1}) = (\phi_{2,2,1} - \phi_{1,2,1}) - (\phi_{2,1,1} - \phi_{1,1,1})$$

Triple differencing can be performed by taking the difference of double differencing performed at time $t_2$ with that performed at time $t_1$. This will eliminate the ambiguity associated with the integral number of wavelengths in carrier phase provided this ambiguity does not change with time. Thus the triple difference result has eliminated all or practically all clock bias errors and the integer ambiguity. Also errors associated with atmospheric delay and satellite ephemeris have been significantly reduced. This triple difference is:

$$\Delta^t(\Delta^r(\Delta^s(\phi_{1,1,1})))$$

Triple difference results can be used to estimate unknown variables. For example, if the position of receiver 1 is known but the position of receiver 2 unknown, it may be possible to estimate the position of receiver 2 using numerical root finding and least squares. Triple difference results for three independent time pairs quite possibly will be sufficient to solve for the three components of position of receiver 2. This may require the use of a numerical procedure such as one of those found in the chapter on root finding and nonlinear sets of equations in Numerical Recipes. To use such a numerical method, an initial approximation of the position of receiver 2 is required. This initial value could probably be provided by a position approximation based on the navigation message and the intersection of sphere surfaces. Although multidimensional numerical root finding can have problems, this disadvantage may be overcome with this good initial estimate. This procedure using three time pairs and a fairly good initial value followed by iteration will result in one observed triple difference result for receiver 2 position. Greater accuracy may be obtained by processing triple difference results for additional sets of three independent time pairs. This will result in an over determined system with multiple solutions. To get estimates for an over determined system, least squares can be used. The least squares procedure determines the position of receiver 2 which best fits the observed triple difference results for receiver 2 positions under the criterion of minimizing the sum of the squares.

# Error Analysis for the Global Positioning System

The analysis of errors computed using the Global Positioning System is important for understanding how GPS works, and for knowing what magnitude of errors should be expected. The Global Positioning System makes corrections for receiver clock errors and other effects but there are still residual errors which are not corrected. The Global Positioning System (GPS) was created by the United States Department of Defense (DOD) in the 1970s. It has come to be widely used for navigation both by the U.S. military and the general public.

Artist's conception of GPS Block II-F satellite in orbit

GPS receiver position is computed based on data received from the satellites. Errors depend on geometric dilution of precision and the sources listed in the table below.

## Overview

| Sources of User Equivalent Range Errors (UERE) | |
|---|---|
| **Source** | **Effect (m)** |
| Signal arrival C/A | ±3 |
| Signal arrival P(Y) | ±0.3 |
| Ionospheric effects | ±5 |
| Ephemeris errors | ±2.5 |
| Satellite clock errors | ±2 |
| Multipath distortion | ±1 |
| Tropospheric effects | ±0.5 |
| C/A | ±6.7 |
| P(Y) | ±6.0 |

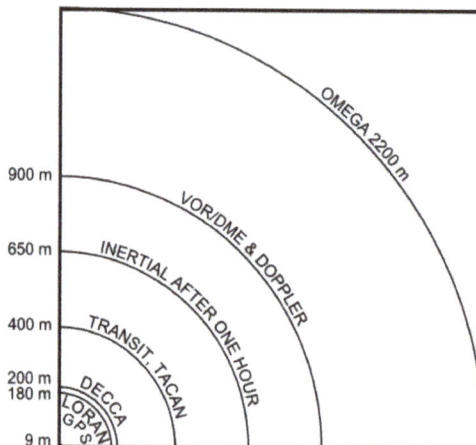

ACCURACY OF NAVIGATION SYSTEMS
(2-dimensional)

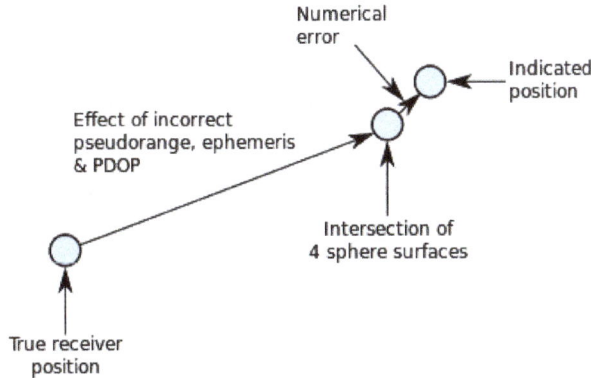

Geometric Error Diagram Showing Typical Relation of Indicated Receiver Position, Intersection of Sphere Surfaces, and True Receiver Position in Terms of Pseudorange Errors, PDOP, and Numerical Errors

User equivalent range errors (UERE) are shown in the table. There is also a numerical error with an estimated value, $\sigma_{num}$, of about 1 meter. The standard deviations, $\sigma_R$, for the coarse/acquisition (C/A) and precise codes are also shown in the table. These standard deviations are computed by taking the square root of the sum of the squares of the individual components (i.e., RSS for root sum squares). To get the standard deviation of receiver position estimate, these range errors must be multiplied by the appropriate dilution of precision terms and then RSS'ed with the numerical error. Electronics errors are one of several accuracy-degrading effects outlined in the table above. When taken together, autonomous civilian GPS horizontal position fixes are typically accurate to about 15 meters (50 ft). These effects also reduce the more precise P(Y) code's accuracy. However, the advancement of technology means that in the present, civilian GPS fixes under a clear view of the sky are on average accurate to about 5 meters (16 ft) horizontally.

The term user equivalent range error (UERE) refers to the error of a component in the distance from receiver to a satellite. These UERE errors are given as ± errors thereby implying that they are unbiased or zero mean errors. These UERE errors are therefore used in computing standard deviations. The standard deviation of the error in receiver position, $\sigma_R$, is computed by multiplying PDOP (Position Dilution Of Precision) by $\sigma_R$, the standard deviation of the user equivalent range errors. $\sigma_R$ is computed by taking the square root of the sum of the squares of the individual component standard deviations.

PDOP is computed as a function of receiver and satellite positions. A detailed description of how to calculate PDOP is given in the section, geometric dilution of precision computation (GDOP).

$\sigma_R$ for the C/A code is given by:

$$3\sigma_R = \sqrt{3^2 + 5^2 + 2.5^2 + 2^2 + 1^2 + 0.5^2}\ \text{m} = 6.7\text{m}$$

The standard deviation of the error in estimated receiver position $\sigma_{rc}$, again for the C/A code is given by:

$$\sigma_{rc} = \sqrt{PDOP^2 \times \sigma_R^2 + \sigma_{num}^2} = \sqrt{PDOP^2 \times 2.2^2 + 1^2}\ m$$

The error diagram on the left shows the inter relationship of indicated receiver position, true receiver position, and the intersection of the four sphere surfaces.

## Signal Arrival Time Measurement

The position calculated by a GPS receiver requires the current time, the position of the satellite and the measured delay of the received signal. The position accuracy is primarily dependent on the satellite position and signal delay.

To measure the delay, the receiver compares the bit sequence received from the satellite with an internally generated version. By comparing the rising and trailing edges of the bit transitions, modern electronics can measure signal offset to within about one percent of a bit pulse width, $\dfrac{0.01}{(1.023 \times 10^6\ /\ s)}$, or approximately 10 nanoseconds for the C/A code. Since GPS signals propagate at the speed of light, this represents an error of about 3 meters.

This component of position accuracy can be improved by a factor of 10 using the higher-chiprate P(Y) signal. Assuming the same one percent of bit pulse width accuracy, the high-frequency P(Y) signal results in an accuracy of $\dfrac{(0.01 \times 300,000,000\ m/s)}{(10.23 \times 10^6\ /\ s)}$ or about 30 centimeters.

## Atmospheric Effects

Inconsistencies of atmospheric conditions affect the speed of the GPS signals as they pass through the Earth's atmosphere, especially the ionosphere. Correcting these errors is a significant challenge to improving GPS position accuracy. These effects are smallest when the satellite is directly overhead and become greater for satellites nearer the horizon since the path through the atmosphere is longer. Once the receiver's approximate location is known, a mathematical model can be used to estimate and compensate for these errors.

Ionospheric delay of a microwave signal depends on its frequency. It arises from ionized atmosphere. This phenomenon is known as dispersion and can be calculated from measurements of delays for two or more frequency bands, allowing delays at other frequencies to be estimated. Some military and expensive survey-grade civilian receivers calculate atmospheric dispersion from the different delays in the L1 and L2 frequencies, and apply a more precise correction. This can be done in civilian receivers without decrypting the P(Y) signal carried on L2, by tracking the carrier wave instead of the modulated code. To facilitate this on lower cost receivers, a new civilian code signal on L2, called L2C, was added to the Block IIR-M satellites, which was first launched in 2005. It allows a direct comparison of the L1 and L2 signals using the coded signal instead of the carrier wave.

The effects of the ionosphere generally change slowly, and can be averaged over time. Those for any particular geographical area can be easily calculated by comparing the GPS-measured position to a known surveyed location. This correction is also valid for other receivers in the same general location. Several systems send this information over radio or other links to allow L1-only receivers to make ionospheric corrections. The ionospheric data are transmitted via satellite in Satellite Based Augmentation Systems (SBAS) such as Wide Area Augmentation System (WAAS) (available in North America and Hawaii), EGNOS (Europe and Asia) or Multi-functional Satellite Augmentation System (MSAS) (Japan), which transmits it on the GPS frequency using a special pseudo-random noise sequence (PRN), so only one receiver and antenna are required.

Humidity also causes a variable delay, resulting in errors similar to ionospheric delay, but occurring in the troposphere. This effect both is more localized and changes more quickly than ionospheric effects, and is not frequency dependent. These traits make precise measurement and compensation of humidity errors more difficult than ionospheric effects.

The Atmospheric pressure can also change the signals reception delay, due to the dry gases present at the troposphere (78% N2, 21% O2, 0.9% Ar...). Its effect varies with local temperature and atmospheric pressure in quite a predictable manner using the laws of the ideal gases.

## Multipath Effects

GPS signals can also be affected by multipath issues, where the radio signals reflect off surrounding terrain; buildings, canyon walls, hard ground, etc. These delayed signals cause measurement errors that are different for each type of GPS signal due to its dependency on the wavelength.

A variety of techniques, most notably narrow correlator spacing, have been developed to mitigate multipath errors. For long delay multipath, the receiver itself can recognize the wayward signal and discard it. To address shorter delay multipath from the signal reflecting off the ground, specialized antennas (e.g., a choke ring antenna) may be used to reduce the signal power as received by the antenna. Short delay reflections are harder to filter out because they interfere with the true signal, causing effects almost indistinguishable from routine fluctuations in atmospheric delay.

Multipath effects are much less severe in moving vehicles. When the GPS antenna is moving, the false solutions using reflected signals quickly fail to converge and only the direct signals result in stable solutions.

## Ephemeris and Clock Errors

While the ephemeris data is transmitted every 30 seconds, the information itself may be up to two hours old. Variability in solar radiation pressure has an indirect effect on GPS accuracy due to its effect on ephemeris errors. If a fast time to first fix (TTFF) is needed, it is possible to upload a valid ephemeris to a receiver, and in addition to setting the time, a position fix can be obtained in under ten seconds. It is feasible to put such ephemeris data on the web so it can be loaded into mobile GPS devices.

The satellite's atomic clocks experience noise and clock drift errors. The navigation message contains corrections for these errors and estimates of the accuracy of the atomic clock. However, they are based on observations and may not indicate the clock's current state.

These problems tend to be very small, but may add up to a few meters (tens of feet) of inaccuracy.

For very precise positioning (e.g., in geodesy), these effects can be eliminated by differential GPS: the simultaneous use of two or more receivers at several survey points. In the 1990s when receivers were quite expensive, some methods of *quasi-differential* GPS were developed, using only *one* receiver but reoccupation of measuring points. At the TU Vienna the method was named *qGPS* and post processing software was developed.

## Geometric Dilution of Precision Computation (GDOP)

## Computation of Geometric Dilution of Precision

The concept of geometric dilution of precision was introduced in the section, *error sources and analysis*. Computations were provided to show how PDOP was used and how it affected the receiver position error standard deviation.

When visible GPS satellites are close together in the sky (i.e., small angular separation), the DOP values are high; when far apart, the DOP values are low. Conceptually, satellites that are close together cannot provide as much information as satellites that are widely separated. Low DOP values represent a better GPS positional accuracy due to the wider angular separation between the satellites used to calculate GPS receiver position. HDOP, VDOP, PDOP and TDOP are respectively Horizontal, Vertical, Position (3-D) and Time Dilution of Precision.

Figure 3.1 Dilution of Precision of Navstar GPS data from the U.S. Coast Guard provide a graphical indication of how geometry affect accuracy.

We now take on the task of how to compute the dilution of precision terms. As a first step in computing DOP, consider the unit vector from the receiver to satellite i with components $\frac{(x_i - x)}{R_i}$, $\frac{(y_i - y)}{R_i}$, and $\frac{(z_i - z)}{R_i}$ where the distance from receiver to the satellite,

$R_i$, , is given by:

$$R_i = \sqrt{(x_i - x)^2 + (y_i - y)^2 + (z_i - z)^2}$$

where $x, y,$ and $z$ denote the position of the receiver and $x_i, y_i,$ and $z_i$ denote the position of satellite $i$. These $x, y,$ and $z$ components may be components in a North, East, Down coordinate system a South, East, Up coordinate system or other convenient system. Formulate the matrix $A$ as:

$$A = \begin{bmatrix} \frac{(x_1 - x)}{R_1} & \frac{(y_1 - y)}{R_1} & \frac{(z_1 - z)}{R_1} & 1 \\ \frac{(x_2 - x)}{R_2} & \frac{(y_2 - y)}{R_2} & \frac{(z_2 - z)}{R_2} & 1 \\ \frac{(x_3 - x)}{R_3} & \frac{(y_3 - y)}{R_3} & \frac{(z_3 - z)}{R_3} & 1 \\ \frac{(x_4 - x)}{R_4} & \frac{(y_4 - y)}{R_4} & \frac{(z_4 - z)}{R_4} & 1 \end{bmatrix}$$

The first three elements of each row of $A$ are the components of a unit vector from the receiver to the indicated satellite. The elements in the fourth column are c where c denotes the speed of light. Formulate the matrix, $Q$, as

$$Q = (A^T A)^{-1}$$

This computation is in accordance with Chapter 11 of The global positioning system by Parkinson and Spilker where the weighting matrix, $P$, has been set to the identity matrix. The elements of the $Q$ matrix are designated as:

$$Q = \begin{bmatrix} d_x^2 & d_{xy}^2 & d_{xz}^2 & d_{xt}^2 \\ d_{xy}^2 & d_y^2 & d_{yz}^2 & d_{yt}^2 \\ d_{xz}^2 & d_{yz}^2 & d_z^2 & d_{zt}^2 \\ d_{xt}^2 & d_{yt}^2 & d_{zt}^2 & d_t^2 \end{bmatrix}$$

The Greek letter $\sigma$ is used quite often where we have used $d$. However the elements of the $Q$ matrix do not represent variances and covariances as they are defined in probability and statistics. Instead they are strictly geometric terms. Therefore, d as in dilution of precision is used. PDOP, TDOP and GDOP are given by

$$PDOP = \sqrt{d_x^2 + d_y^2 + d_z^2}$$

$$TDOP = \sqrt{d_t^2} = |d_t|$$

$$GDOP = \sqrt{PDOP^2 + TDOP^2}$$

in agreement with "Section 1.4.9 of PRINCIPLES OF SATELLITE POSITIONING".

The horizontal dilution of precision, $HDOP = \sqrt{d_x^2 + d_y^2}$, and the vertical dilution of precision, $VDOP = \sqrt{d_z^2} = |d_z|$, are both dependent on the coordinate system used. To correspond to the local horizon plane and the local vertical, $x$, $y$, and $z$ should denote positions in either a North, East, Down coordinate system or a South, East, Up coordinate system.

## Derivation of Equations for Computing Geometric Dilution of Precision

The equations for computing the geometric dilution of precision terms have been described in the previous section. This section describes the derivation of these equations. The method used here is similar to that used in "Global Positioning System (preview) by Parkinson and Spiker"

Consider the position error vector, $\mathbf{e}$, defined as the vector from the intersection of the four sphere surfaces corresponding to the pseudoranges to the true position of the receiver. $\mathbf{e} = e_x \hat{x} + e_y \hat{y} + e_z \hat{z}$ where bold denotes a vector and $\hat{x}$, $\hat{y}$, and $\hat{z}$ denote unit vectors along the x, y, and z axes respectively. Let $e_t$ denote the time error, the true time minus the receiver indicated time. Assume that the mean value of the three components of $\mathbf{e}$ and $e_t$ are zero.

$$A \begin{bmatrix} e_x \\ e_y \\ e_z \\ e_t \end{bmatrix} = \begin{bmatrix} \dfrac{(x_1-x)}{R_1} & \dfrac{(y_1-y)}{R_1} & \dfrac{(z_1-z)}{R_1} & 1 \\ \dfrac{(x_2-x)}{R_2} & \dfrac{(y_2-y)}{R_2} & \dfrac{(z_2-z)}{R_2} & 1 \\ \dfrac{(x_3-x)}{R_3} & \dfrac{(y_3-y)}{R_3} & \dfrac{(z_3-z)}{R_3} & 1 \\ \dfrac{(x_4-x)}{R_4} & \dfrac{(y_4-y)}{R_4} & \dfrac{(z_4-z)}{R_4} & 1 \end{bmatrix} \begin{bmatrix} e_x \\ e_y \\ e_z \\ e_t \end{bmatrix} = \begin{bmatrix} e_1 \\ e_2 \\ e_3 \\ e_4 \end{bmatrix} \quad (1)$$

where $e_1$, $e_2$, $e_3$, and $e_4$ are the errors in pseudoranges 1 through 4 respectively. This equation comes from linearizing the Newton-Raphson equation relating pseudoranges to receiver position, satellite positions, and receiver clock errors. Multiplying both sides by $A^{-1}$ there results

$$\begin{bmatrix} e_x \\ e_y \\ e_z \\ e_t \end{bmatrix} = A^{-1} \begin{bmatrix} e_1 \\ e_2 \\ e_3 \\ e_4 \end{bmatrix} \quad (2)$$

Transposing both sides:

$$\begin{bmatrix} e_x & e_y & e_z & e_t \end{bmatrix} = \begin{bmatrix} e_1 & e_2 & e_3 & e_4 \end{bmatrix} \left( A^{-1} \right)^T \quad (3).$$

Post multiplying the matrices on both sides of equation (2) by the corresponding matrices in equation (3), there results

$$\begin{bmatrix} e_x \\ e_y \\ e_z \\ e_t \end{bmatrix} \begin{bmatrix} e_x & e_y & e_z & e_t \end{bmatrix} = A^{-1} \begin{bmatrix} e_1 \\ e_2 \\ e_3 \\ e_4 \end{bmatrix} \begin{bmatrix} e_1 & e_2 & e_3 & e_4 \end{bmatrix} \left( A^{-1} \right)^T \quad (4).$$

Taking the expected value of both sides and taking the non-random matrices outside the expectation operator, E, there results:

$$E \left( \begin{bmatrix} e_x \\ e_y \\ e_z \\ e_t \end{bmatrix} \begin{bmatrix} e_x & e_y & e_z & e_t \end{bmatrix} \right) = A^{-1} E \left( \begin{bmatrix} e_1 \\ e_2 \\ e_3 \\ e_4 \end{bmatrix} \begin{bmatrix} e_1 & e_2 & e_3 & e_4 \end{bmatrix} \right) \left( A^{-1} \right)^T \quad (5)$$

Assuming the pseudorange errors are uncorrelated and have the same variance, the co-variance matrix on the right side can be expressed as a scalar times the identity matrix. Thus

$$\begin{bmatrix} \sigma_x^2 & \sigma_{xy}^2 & \sigma_{xz}^2 & \sigma_{xt}^2 \\ \sigma_{xy}^2 & \sigma_y^2 & \sigma_{yz}^2 & \sigma_{yt}^2 \\ \sigma_{xz}^2 & \sigma_{yz}^2 & \sigma_z^2 & \sigma_{zt}^2 \\ \sigma_{xt}^2 & \sigma_{yt}^2 & \sigma_{zt}^2 & \sigma_t^2 \end{bmatrix} = \sigma_R^2 \, A^{-1} \left( A^{-1} \right)^T = \sigma_R^2 \left( A^T A \right)^{-1} \quad (6)$$

since $A^{-1} \left( A^{-1} \right)^T \left( A^T A \right) = I$

Note: $\left( A^{-1} \right)^T = \left( A^T \right)^{-1}$, since $I = \left( AA^{-1} \right)^T = \left( A^{-1} \right)^T A^T$

Substituting for $\left( A^T A \right)^{-1} = Q$ there follows

$$\begin{bmatrix} \sigma_x^2 & \sigma_{xy}^2 & \sigma_{xz}^2 & \sigma_{xt}^2 \\ \sigma_{xy}^2 & \sigma_y^2 & \sigma_{yz}^2 & \sigma_{yt}^2 \\ \sigma_{xz}^2 & \sigma_{yz}^2 & \sigma_z^2 & \sigma_{zt}^2 \\ \sigma_{xt}^2 & \sigma_{yt}^2 & \sigma_{zt}^2 & \sigma_t^2 \end{bmatrix} = \sigma_R^2 \begin{bmatrix} d_x^2 & d_{xy}^2 & d_{xz}^2 & d_{xt}^2 \\ d_{xy}^2 & d_y^2 & d_{yz}^2 & d_{yt}^2 \\ d_{xz}^2 & d_{yz}^2 & d_z^2 & d_{zt}^2 \\ d_{xt}^2 & d_{yt}^2 & d_{zt}^2 & d_t^2 \end{bmatrix} \quad (7)$$

From equation (7), it follows that the variances of indicated receiver position and time are

$$\sigma_{rc}^2 = \sigma_x^2 + \sigma_y^2 + \sigma_z^2 = \sigma_R^2 \left( d_x^2 + d_y^2 + d_z^2 \right) = PDOP^2 \sigma_R^2 \text{ and}$$

$$\sigma_t^2 = \sigma_R^2 d_t^2 = TDOP^2 \sigma_R^2$$

The remaining position and time error variance terms follow in a straightforward manner.

## Selective Availability

GPS included a (currently disabled) feature called *Selective Availability* (*SA*) that adds intentional, time varying errors of up to 100 meters (328 ft) to the publicly available navigation signals. This was intended to deny an enemy the use of civilian GPS receivers for precision weapon guidance.

SA errors are actually pseudorandom, generated by a cryptographic algorithm from a classified *seed* key available only to authorized users (the U.S. military, its allies and a few other users, mostly government) with a special military GPS receiver. Mere possession of the receiver is insufficient; it still needs the tightly controlled daily key.

Before it was turned off on May 2, 2000, typical SA errors were about 50 m (164 ft) horizontally and about 100 m (328 ft) vertically. Because SA affects every GPS receiver in a given area almost equally, a fixed station with an accurately known position can measure the SA error values and transmit them to the local GPS receivers so they may correct their position fixes. This is called Differential GPS or *DGPS*. DGPS also corrects for several other important sources of GPS errors, particularly ionospheric delay, so it continues to be widely used even though SA has been turned off. The ineffectiveness of SA in the face of widely available DGPS was a common argument for turning off SA, and this was finally done by order of President Clinton in 2000.

DGPS services are widely available from both commercial and government sources. The latter include WAAS and the U.S. Coast Guard's network of LF marine navigation beacons. The accuracy of the corrections depends on the distance between the user and the DGPS receiver. As the distance increases, the errors at the two sites will not correlate as well, resulting in less precise differential corrections.

During the 1990–91 Gulf War, the shortage of military GPS units caused many troops and their families to buy readily available civilian units. Selective Availability significantly impeded the U.S. military's own battlefield use of these GPS, so the military made the decision to turn it off for the duration of the war.

In the 1990s, the FAA started pressuring the military to turn off SA permanently. This would save the FAA millions of dollars every year in maintenance of their own radio navigation systems. The amount of error added was "set to zero" at midnight on May 1, 2000 following an announcement by U.S. President Bill Clinton, allowing users access to the error-free L1 signal. Per the directive, the induced error of SA was changed to add no error to the public signals (C/A code). Clinton's executive order required SA to be set to zero by 2006; it happened in 2000 once the U.S. military developed a new system that provides the ability to deny GPS (and other navigation services) to hostile forces in a specific area of crisis without affecting the rest of the world or its own military systems.

On 19 September 2007, the United States Department of Defense announced that future GPS III satellites will not be capable of implementing SA, eventually making the policy permanent.

## Anti-spoofing

Another restriction on GPS, antispoofing, remains on. This encrypts the *P-code* so that it cannot be mimicked by a transmitter sending false information. Few civilian receivers have ever used the P-code, and the accuracy attainable with the public C/A code was much better than originally expected (especially with DGPS). So much so that the antispoof policy has relatively little effect on most civilian users. Turning off antispoof would primarily benefit surveyors and some scientists who need extremely precise positions for experiments such as tracking tectonic plate motion.

## Relativity

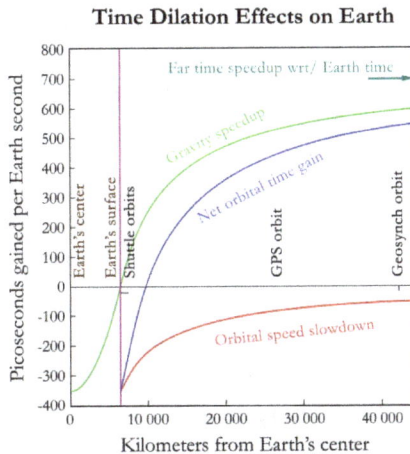

Satellite clocks are slowed by their orbital speed but sped up by their distance out of the Earth's gravitational well.

A number of sources of error exist due to relativistic effects that would render the system useless if uncorrected. Three relativistic effects are the time dilation, gravitational frequency shift, and eccentricity effects. Examples include the relativistic time *slowing* due to the speed of the satellite of about 1 part in $10^{10}$, the gravitational time dilation that makes a satellite run about 5 parts in $10^{10}$ *faster* than an Earth-based clock, and the Sagnac effect due to rotation relative to receivers on Earth. These topics are examined below, one at a time.

## Special and General Relativity

According to the theory of relativity, due to their constant movement and height relative to the Earth-centered, non-rotating approximately inertial reference frame, the clocks on the satellites are affected by their speed. Special relativity predicts that the frequency of the atomic clocks moving at GPS orbital speeds will tick more slowly than

stationary ground clocks by a factor of $\dfrac{v^2}{2c^2} \approx 10^{-10}$, or result in a delay of about 7 μs/ day, where the orbital velocity is v = 4 km/s, and c = the speed of light. The time dilation effect has been measured and verified using the GPS.

The effect of gravitational frequency shift on the GPS due to general relativity is that a clock closer to a massive object will be slower than a clock farther away. Applied to the GPS, the receivers are much closer to Earth than the satellites, causing the GPS clocks to be faster by a factor of 5×10^(−10), or about 45.9 μs/day. This gravitational frequency shift is noticeable.

When combining the time dilation and gravitational frequency shift, the discrepancy is about 38 microseconds per day, a difference of 4.465 parts in $10^{10}$. Without correction, errors in the initial pseudorange of roughly 10 km/day would accumulate. This initial pseudorange error is corrected in the process of solving the navigation equations. In addition the elliptical, rather than perfectly circular, satellite orbits cause the time dilation and gravitational frequency shift effects to vary with time. This eccentricity effect causes the clock rate difference between a GPS satellite and a receiver to increase or decrease depending on the altitude of the satellite.

To compensate for the discrepancy, the frequency standard on board each satellite is given a rate offset prior to launch, making it run slightly slower than the desired frequency on Earth; specifically, at 10.22999999543 MHz instead of 10.23 MHz. Since the atomic clocks on board the GPS satellites are precisely tuned, it makes the system a practical engineering application of the scientific theory of relativity in a real-world environment. Placing atomic clocks on artificial satellites to test Einstein's general theory was proposed by Friedwardt Winterberg in 1955.

## Calculation of Time Dilation

To calculate the amount of daily time dilation experienced by GPS satellites relative to Earth we need to separately determine the amounts due to special relativity (velocity) and general relativity (gravity) and add them together.

The amount due to velocity will be determined using the Lorentz transformation. This will be:

$$\frac{1}{\gamma} = \sqrt{1 - \frac{v^2}{c^2}}$$

For small values of $v/c$, by using binomial expansion this approximates to:

$$\frac{1}{\gamma} \approx 1 - \frac{v^2}{2c^2}$$

The GPS satellites move at 3874 m/s relative to Earth's center. We thus determine:

$$\frac{1}{\gamma} \approx 1 - \frac{3874^2}{2\left(2.998 \times 10^8\right)^2} \approx 1 - 8.349 \times 10^{-11}$$

This difference below 1 of $8.349 \times 10^{-11}$ represents the fraction by which the satellites' clocks move slower than Earth's. It is then multiplied by the number of nanoseconds in a day:

$$-8.349 \times 10^{-11} \times 60 \times 60 \times 24 \times 10^9 \approx -7214 \text{ ns}$$

That is, the satellites' clocks lose 7,214 nanoseconds a day due to special relativity effects.

> Note that this speed of 3874 m/s is measured relative to Earth's center rather than its surface where the GPS receivers (and users) are. This is because Earth's equipotential makes net time dilation equal across its geodesic surface. That is, the combination of Special and General effects make the net time dilation at the equator equal to that of the poles, which in turn are at rest relative to the center. Hence we use the center as a reference point to represent the entire surface.

The amount of dilation due to gravity will be determined using the gravitational time dilation equation:

$$\frac{1}{\gamma} = \sqrt{1 - \frac{2GM}{rc^2}}$$

For small values of $M/r$, by using binomial expansion this approximates to:

$$\frac{1}{\gamma} \approx 1 - \frac{GM}{rc^2}$$

We are again only interested in the fraction below 1, and in the difference between Earth and the satellites. To determine this difference we take:

$$\Delta\left(\frac{1}{\gamma}\right) \approx \frac{GM_{earth}}{R_{earth}c^2} - \frac{GM_{earth}}{R_{gps}c^2}$$

Earth has a radius of 6,357 km (at the poles) making $R_{earth}$ = 6,357,000 m and the satellites have an altitude of 20,184 km making their orbit radius $R_{gps}$ = 26,541,000 m. Substituting these in the above equation, with $M_{earth}$ = 5.974×10²⁴, $G$ = 6.674×10⁻¹¹, and $c$ = 2.998×10⁸ (all in SI units), gives:

$$\Delta\left(\frac{1}{\gamma}\right) \approx 5.307 \times 10^{-10}$$

This represents the fraction by which the satellites' clocks move faster than Earth's. It is then multiplied by the number of nanoseconds in a day:

$$5.307 \times 10^{-10} \times 60 \times 60 \times 24 \times 10^9 \approx 45850 \text{ ns}$$

That is, the satellites' clocks gain 45,850 nanoseconds a day due to general relativity effects. These effects are added together to give (rounded to 10 ns):

$$45850 - 7210 = 38640 \text{ ns}$$

Hence the satellites' clocks gain approximately 38,640 nanoseconds a day or 38.6 μs per day due to relativity effects in total.

In order to compensate for this gain, a GPS clock's frequency needs to be slowed by the fraction:

$$5.307 \times 10^{-10} - 8.349 \times 10^{-11} = 4.472 \times 10^{-10}$$

This fraction is subtracted from 1 and multiplied by the pre-adjusted clock frequency of 10.23 MHz:

$$(1 - 4.472 \times 10^{-10}) \times 10.23 = 10.22999999543$$

That is, we need to slow the clocks down from 10.23 MHz to 10.22999999543 MHz in order to negate the effects of relativity.

## Sagnac Distortion

GPS observation processing must also compensate for the Sagnac effect. The GPS time scale is defined in an inertial system but observations are processed in an Earth-centered, Earth-fixed (co-rotating) system, a system in which simultaneity is not uniquely defined. A coordinate transformation is thus applied to convert from the inertial system to the ECEF system. The resulting signal run time correction has opposite algebraic signs for satellites in the Eastern and Western celestial hemispheres. Ignoring this effect will produce an east–west error on the order of hundreds of nanoseconds, or tens of meters in position.

## Natural Sources of Interference

Since GPS signals at terrestrial receivers tend to be relatively weak, natural radio signals or scattering of the GPS signals can desensitize the receiver, making acquiring and tracking the satellite signals difficult or impossible.

Space weather degrades GPS operation in two ways, direct interference by solar radio burst noise in the same frequency band or by scattering of the GPS radio signal in ionospheric irregularities referred to as scintillation. Both forms of degradation follow the 11 year solar cycle and are a maximum at sunspot maximum although they can occur at any time. Solar radio bursts are associated with solar flares and coronal mass ejections (CMEs) and their impact can affect reception over the half of the Earth facing the sun.

Scintillation occurs most frequently at tropical latitudes where it is a night time phenomenon. It occurs less frequently at high latitudes or mid-latitudes where magnetic storms can lead to scintillation. In addition to producing scintillation, magnetic storms can produce strong ionospheric gradients that degrade the accuracy of SBAS systems.

## Artificial Sources of Interference

In automotive GPS receivers, metallic features in windshields, such as defrosters, or car window tinting films can act as a Faraday cage, degrading reception just inside the car.

Man-made EMI (electromagnetic interference) can also disrupt or jam GPS signals. In one well-documented case it was impossible to receive GPS signals in the entire harbor of Moss Landing, California due to unintentional jamming caused by malfunctioning TV antenna preamplifiers. Intentional jamming is also possible. Generally, stronger signals can interfere with GPS receivers when they are within radio range or line of sight. In 2002 a detailed description of how to build a short-range GPS L1 C/A jammer was published in the online magazine Phrack.

The U.S. government believes that such jammers were used occasionally during the War in Afghanistan, and the U.S. military claims to have destroyed six GPS jammers during the Iraq War, including one that was destroyed with a GPS-guided bomb. A GPS jammer is relatively easy to detect and locate, making it an attractive target for anti-radiation missiles. The UK Ministry of Defence tested a jamming system in the UK's West Country on 7 and 8 June 2007.

Some countries allow the use of GPS repeaters to allow the reception of GPS signals indoors and in obscured locations; while in other countries these are prohibited as the retransmitted signals can cause multi-path interference to other GPS receivers that receive data from both GPS satellites and the repeater. In the UK Ofcom now permits the use of GPS/GNSS Repeaters under a 'light licensing' regime.

Due to the potential for both natural and man-made noise, numerous techniques continue to be developed to deal with the interference. The first is to not rely on GPS as a sole source. According to John Ruley, "IFR pilots should have a fallback plan in case of a GPS malfunction". Receiver Autonomous Integrity Monitoring (RAIM) is a feature included in some receivers, designed to provide a warning to the user if jamming or another problem is detected. The U.S. military has also deployed since 2004 their Selective Availability / Anti-Spoofing Module (SAASM) in the Defense Advanced GPS Receiver (DAGR). In demonstration videos the DAGR was shown to detect jamming and maintain its lock on the encrypted GPS signals during interference which caused civilian receivers to lose lock.

## References

- O'Leary, Beth Laura; Darrin, Ann Garrison (2009). Handbook of Space Engineering, Archaeology, and Heritage. Hoboken: CRC Press. pp. 239–240. ISBN 9781420084320.

- Michael Russell Rip; James M. Hasik (2002). The Precision Revolution: GPS and the Future of Aerial Warfare. Naval Institute Press. p. 65. ISBN 1-55750-973-5. Retrieved January 14, 2010.

- Dietrich Schroeer; Mirco Elena (2000). Technology Transfer. Ashgate. p. 80. ISBN 0-7546-2045-X. Retrieved May 25, 2008.

- Michael Russell Rip; James M. Hasik (2002). The Precision Revolution: GPS and the Future of Aerial Warfare. Naval Institute Press. ISBN 1-55750-973-5. Retrieved May 25, 2008.

- Grewal, Mohinder S.; Weill, Lawrence Randolph; Andrews, Angus P. (2001). Global positioning systems, inertial navigation, and integration. John Wiley and Sons. ISBN 978-0-47135-032-3.

- Parkinson; Spilker (1996). The global positioning system. American Institute of Aeronautics & Astronomy. ISBN 978-1-56347-106-3.

- Webb, Stephen (2004). Out of this world: colliding universes, branes, strings, and other wild ideas of modern physics. Springer. ISBN 0-387-02930-3. Retrieved 2013-08-16.

- "2011 John Deere StarFire 3000 Operator Manual" (PDF). John Deere. Archived from the original (PDF) on January 5, 2012. Retrieved November 13, 2011.

- "Federal Communications Commission Report and Order In the Matter of Fixed and Mobile Services in the Mobile Satellite Service Bands at 1525–1559 MHz and 1626.5–1660.5 MHz" (PDF). FCC.gov. April 6, 2011. Retrieved December 13, 2011.

- "Federal Communications Commission Table of Frequency Allocations" (PDF). FCC.gov. November 18, 2011. Retrieved December 13, 2011.

- "Statement of Julius P. Knapp, Chief, Office of Engineering and Technology, Federal Communications Commission" (PDF). gps.gov. September 15, 2011. p. 3. Retrieved December 13, 2011.

- "FCC Order, Granted LightSquared Subsidiary LLC, a Mobile Satellite Service licensee in the L-Band, a conditional waiver of the Ancillary Terrestrial Component "integrated service" rule" (PDF). Federal Communications Commission. FCC.Gov. January 26, 2011. Retrieved December 13, 2011.

# Methods and uses of Satellite Navigation System

Satellite navigation systems are used in a variety of ways and this chapter focuses on vehicle tracking systems and automated vehicle locations. The content delves into the types of systems used in these applications and the conventional and unconventional uses of each. A section also talks about an important aspect of GPS systems namely, dilution of precision and its components.

## Dilution of Precision (Navigation)

Dilution of precision (DOP), or geometric dilution of precision (GDOP), is a term used in satellite navigation and geomatics engineering to specify the additional multiplicative effect of navigation satellite geometry on positional measurement precision.

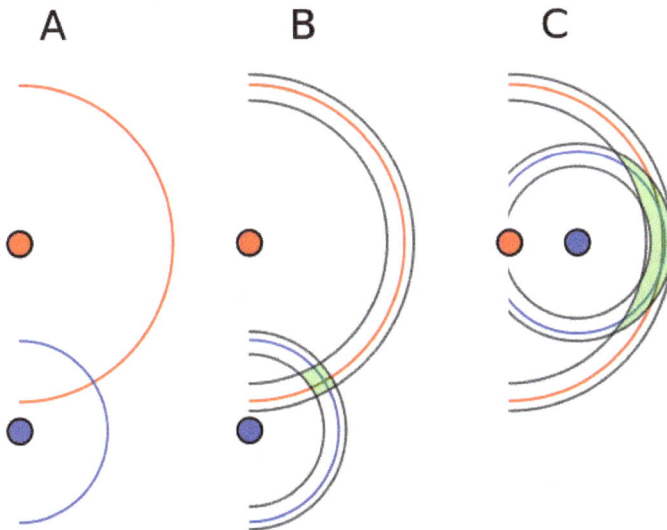

Understanding the Geometric Dilution of Precision (GDOP) with a simple example. In A someone has measured the distance to two landmarks, and plotted their point as the intersection of two circles with the measured radius. In B the measurement has some error bounds, and their true location will lie anywhere in the green area. In C the measurement error is the same, but the error on their position has grown considerably due to the arrangement of the landmarks.

Navigation satellites with poor geometry for Geometric Dilution of Precision (GDOP).

Navigation satellites with good geometry for Geometric Dilution of Precision (GDOP).

## Introduction

The concept of dilution of precision (DOP) originated with users of the Loran-C navigation system. The idea of Geometric DOP is to state how errors in the measurement will affect the final state estimation. This can be defined as:

$$GDOP = \frac{\Delta(\text{Output Location})}{\Delta(\text{Measured Data})}$$

Conceptually you can imagine errors on a measurement resulting in the $\Delta$(Measured Data) term changing. Ideally small changes in the measured data will not result in large changes in output location, as such a result would indicate the solution is very sensitive to errors. The interpretation of this formula is shown in the figure to the right, showing two possible scenarios with acceptable and poor GDOP.

More recently, the term has come into much wider usage with the development and adoption of GPS. Neglecting ionospheric and tropospheric effects, the signal from navigation satellites has a fixed precision. Therefore, the relative satellite-receiver geometry plays a major role in determining the precision of estimated positions and times. Due to the relative geometry of any given satellite to a receiver, the precision in the pseudorange of the satellite translates to a corresponding component in each of the four dimensions of position measured by the receiver (i.e., $x$, $y$, $z$, and $t$). The precision of multiple satellites in view of a receiver combine according to the relative position of the satellites to determine the level of precision in each dimension of the receiver measurement. When visible navigation satellites are close together in the sky, the geometry is said to be weak and the DOP value is high; when far apart, the geometry is strong and the DOP value is low. Consider two overlapping rings, or annuli, of different centres. If they overlap at right angles, the greatest extent of the overlap is much smaller than if they overlap in near parallel. Thus a low DOP value represents a better positional precision due to the wider angular separation between the satellites used to calculate a unit's position. Other factors that can increase the effective DOP are obstructions such as nearby mountains or buildings.

DOP can be expressed as a number of separate measurements:

- HDOP – horizontal dilution of precision

- VDOP – vertical dilution of precision

- PDOP – position (3D) dilution of precision

- TDOP – time dilution of precision

These values follow mathematically from the positions of the usable satellites. Signal receivers allow the display of these positions (*skyplot*) as well as the DOP values.

The term can also be applied to other location systems that employ several geographical spaced sites. It can occur in electronic-counter-counter-measures (electronic warfare) when computing the location of enemy emitters (radar jammers and radio communications devices). Using such an interferometry technique can provide certain geometric layout where there are degrees of freedom that cannot be accounted for due to inadequate configurations.

The effect of geometry of the satellites on position error is called geometric dilution of precision and it is roughly interpreted as ratio of position error to the range error. Imagine that a square pyramid is formed by lines joining four satellites with the receiver at the tip of the pyramid. The larger the volume of the pyramid, the better (lower) the value of GDOP; the smaller its volume, the worse (higher) the value of GDOP will be. Similarly, the greater the number of satellites, the better the value of GDOP.

## Meaning of DOP Values

| DOP Value | Rating | Description |
|-----------|--------|-------------|
| < 1 | Ideal | Highest possible confidence level to be used for applications demanding the highest possible precision at all times. |

| 1-2 | Excellent | At this confidence level, positional measurements are considered accurate enough to meet all but the most sensitive applications. |
|---|---|---|
| 2-5 | Good | Represents a level that marks the minimum appropriate for making business decisions. Positional measurements could be used to make reliable in-route navigation suggestions to the user. |
| 5-10 | Moderate | Positional measurements could be used for calculations, but the fix quality could still be improved. A more open view of the sky is recommended. |
| 10-20 | Fair | Represents a low confidence level. Positional measurements should be discarded or used only to indicate a very rough estimate of the current location. |
| >20 | Poor | At this level, measurements are inaccurate by as much as 300 meters with a 6-meter accurate device (50 DOP × 6 meters) and should be discarded. |

The DOP factors are functions of the diagonal elements of the covariance matrix of the parameters, expressed either in a global or a local geodetic frame.

## Computation of DOP Values

As a first step in computing DOP, consider the unit vectors from the receiver to satellite i: $\left( \dfrac{(x_i - x)}{R_i}, \dfrac{(y_i - y)}{R_i}, \dfrac{(z_i - z)}{R_i} \right)$ where $R_i = \sqrt{(x_i - x)^2 + (y_i - y)^2 + (z_i - z)^2}$ and where $x, y$ and $z$ denote the position of the receiver and $x_i, y_i$ and $z_i$ denote the position of satellite i. Formulate the matrix, A, which (for 4 range measurement residual equations) is:

$$A = \begin{bmatrix} \dfrac{(x_1 - x)}{R_1} & \dfrac{(y_1 - y)}{R_1} & \dfrac{(z_1 - z)}{R_1} & -1 \\[2mm] \dfrac{(x_2 - x)}{R_2} & \dfrac{(y_2 - y)}{R_2} & \dfrac{(z_2 - z)}{R_2} & -1 \\[2mm] \dfrac{(x_3 - x)}{R_3} & \dfrac{(y_3 - y)}{R_3} & \dfrac{(z_3 - z)}{R_3} & -1 \\[2mm] \dfrac{(x_4 - x)}{R_4} & \dfrac{(y_4 - y)}{R_4} & \dfrac{(z_4 - z)}{R_4} & -1 \end{bmatrix}$$

The first three elements of each row of $A$ are the components of a unit vector from the receiver to the indicated satellite. If the elements in the fourth column are c which denotes the speed of light then the $\sigma_t$ factor (time dilution) is always 1. If the elements in the fourth column are -1 then the $\sigma_t$ factor is calculated properly. Formulate the matrix, Q, as:

$$Q = \left( A^T A \right)^{-1}$$

In general: $Q = (J_x^T (J_d C_d J_d^T)^{-1} J_x)^{-1}$ where $J_x$ is the Jacobian of the sensor measurement residual equations $f_i(\underline{x}, \underline{d}) = 0$, with respect to the unknowns, $\underline{x}$; $J_d$ is the Jacobian of the sensor measurement residual equations with respect to the measured quantities $\underline{d}$, and $C_d$ is the correlation matrix for noise in the measured quantities. For the preceding case of 4 range measurement residual equations:

$$\underline{x} = (x, y, z, \tau)^T, \quad \underline{d} = (\tau_1, \tau_2, \tau_3, \tau_4)^T, \quad \tau = ct, \quad \tau_i = ct_i, \quad R_i = |\tau_i - \tau| = \sqrt{(\tau_i - \tau)^2},$$

$$f_i(\underline{x}, \underline{d}) = \sqrt{(x_i - x)^2 + (y_i - y)^2 + (z_i - z)^2} - \sqrt{(\tau_i - \tau)^2}, \quad J_x = A, \quad J_d = -I$$ and the measurement noises for the different $\tau_i$ have been assumed to be independent which makes $C_d = I$. This formula for Q arises from applying Best Linear Unbiased Estimation to a linearized version of the sensor measurement residual equations about the current solution $\Delta \underline{x} = -Q * (J_x (J_d C_d J_d^T)^{-1} f)$, except in the case of B.L.U.E. $C_d$ is a noise covariance matrix rather than the noise correlation matrix used in DOP, and the reason DOP makes this substitution is to obtain a relative error. When is a noise covariance matrix, $Q$ is an estimate of the matrix of covariance of noise in the unknowns due to the noise in the measured quantities. It is the estimate obtained by the First ORder second Moment (F.OR.M.) uncertainty quantification technique which was state of the art in the 1980's. In order for the F.OR.M. theory to be strictly applicable, either the input noise distributions need to be Gaussian or the measurement noise standard deviations need to be small relative to rate of change in the output near the solution. In this context, the second criteria is typically the one that is satisfied.

This (i.e. for the 4 range measurement residual equations) computation is in accordance with where the weighting matrix, $P = (J_d C_d J_d^T)^{-1}$, has been set to the identity matrix.

The elements of $Q$ are designated as:

$$Q = \begin{bmatrix} \sigma_x^2 & \sigma_{xy} & \sigma_{xz} & \sigma_{xt} \\ \sigma_{xy} & \sigma_y^2 & \sigma_{yz} & \sigma_{yt} \\ \sigma_{xz} & \sigma_{yz} & \sigma_z^2 & \sigma_{zt} \\ \sigma_{xt} & \sigma_{yt} & \sigma_{zt} & \sigma_t^2 \end{bmatrix}$$

PDOP, TDOP and GDOP are given by:

$$PDOP = \sqrt{\sigma_x^2 + \sigma_y^2 + \sigma_z^2}$$

$$TDOP = \sqrt{\sigma_t^2}$$

$$GDOP = \sqrt{PDOP^2 + TDOP^2}$$

in agreement with Section 1.4.9 of *Principles of Satellite Positioning*. More generally, GDOP is the square root of the trace of the $Q$ matrix.

The horizontal dilution of precision, $HDOP = \sqrt{\sigma_x^2 + \sigma_y^2}$, and the vertical dilution of precision, $VDOP = \sqrt{\sigma_z^2}$, are both dependent on the coordinate system used. To correspond to the local horizon plane and the local vertical, $x$, $y$, and $z$ should denote positions in either a north, east, down coordinate system or a east, north, up coordinate system.

# Vehicle Tracking System

A vehicle tracking system combines the use of automatic vehicle location in individual vehicles with software that collects these fleet data for a comprehensive picture of vehicle locations. Modern vehicle tracking systems commonly use GPS or GLONASS technology for locating the vehicle, but other types of automatic vehicle location technology can also be used. Vehicle information can be viewed on electronic maps via the Internet or specialized software. Urban public transit authorities are an increasingly common user of vehicle tracking systems, particularly in large cities.

## Active Versus Passive Tracking

Several types of vehicle tracking devices exist. Typically they are classified as "passive" and "active". "Passive" devices store GPS location, speed, heading and sometimes a trigger event such as key on/off, door open/closed. Once the vehicle returns to a predetermined point, the device is removed and the data downloaded to a computer for evaluation. Passive systems include auto download type that transfer data via wireless download. "Active" devices also collect the same information but usually transmit the data in near-real-time via cellular or satellite networks to a computer or data center for evaluation.

Many modern vehicle tracking devices combine both active and passive tracking abilities: when a cellular network is available and a tracking device is connected it transmits data to a server; when a network is not available the device stores data in internal memory and will transmit stored data to the server later when the network becomes available again.

Historically, vehicle tracking has been accomplished by installing a box into the vehicle, either self-powered with a battery or wired into the vehicle's power system. For detailed vehicle locating and tracking this is still the predominant method; however, many companies are increasingly interested in the emerging cell phone technologies that provide tracking of multiple entities, such as both a salesperson and their vehicle. These systems also offer tracking of calls, texts, web use and generally provide a wider range of options.

## Typical Architecture

Major constituents of the GPS based tracking are:

1. GPS tracking: The device fits into the vehicle and captures the GPS location information apart from other vehicle information at regular intervals to a central server. Other vehicle information can include fuel amount, engine temperature, altitude, reverse geocoding, door open/close, tire pressure, cut off fuel, turn off ignition, turn on headlight, turn on taillight, battery status, GSM area code/cell code decoded, number of GPS satellites in view, glass open/close, fuel amount, emergency button status, cumulative idling, computed odometer, engine RPM, throttle position, GPRS status and a lot more. Capability of these devices actually decide the final capability of the whole tracking system; most vehicle tracking systems, in addition to providing the vehicle's location data, feature a wide range of communication ports that can be used to integrate other on board systems, allowing to check their status and control or automate their operation.

2. GPS tracking server: The tracking server has three responsibilities: receiving data from the GPS tracking unit, securely storing it, and serving this information on demand to the user.

3. User interface: The UI determines how one will be able to access information, view vehicle data, and elicit important details from it.

## Common Uses

Vehicle tracking systems are commonly used by fleet operators for fleet management functions such as fleet tracking, routing, dispatching, on-board information and security. Along with commercial fleet operators, urban transit agencies use the technology for a number of purposes, including monitoring schedule adherence of buses in service, triggering changes of buses' destination sign displays at the end of the line (or other set location along a bus route), and triggering pre-recorded announcements for passengers.

The American Public Transportation Association estimated that, at the beginning of 2009, around half of all transit buses in the United States were already using a GPS-based vehicle tracking system to trigger automated stop announcements. This can refer to external announcements (triggered by the opening of the bus's door) at a bus stop, announcing the vehicle's route number and destination, primarily for the benefit of visually impaired customers, or to internal announcements (to passengers already on board) identifying the next stop, as the bus (or tram) approaches a stop, or both. Data collected as a transit vehicle follows its route is often continuously fed into a computer program which compares the vehicle's actual location and time with its schedule, and in turn produces a frequently updating display for the driver, telling him/her how early or late he/she is at any given time, potentially making it easier to adhere more closely to the published schedule.

Such programs are also used to provide customers with real-time information as to the waiting time until arrival of the next bus or tram/streetcar at a given stop, based on the nearest vehicles' actual progress at the time, rather than merely giving information as to the *scheduled* time of the next arrival. Transit systems providing this kind of information assign a unique number to each stop, and waiting passengers can obtain information by entering the stop number into an automated telephone system or an application on the transit system's website.

Some transit agencies provide a virtual map on their website, with icons depicting the current locations of buses in service on each route, for customers' information, while others provide such information only to dispatchers or other employees.

With the GPS technology being enhanced day by day, companies are coming up with devices that are compatible with phones and other modern gadgets. These devices provide live time activity of the fleet on personal devices without even logging onto their website as well. Keeping a track of fleet commander's actionable data, improving efficiency, reducing fuel cost etc. also come under fleet tracking. These devices and software help in cost cutting.

Other applications include monitoring driving behavior, such as an employer of an employee, or a parent with a teen driver.

Vehicle tracking systems are also popular in consumer vehicles as a theft prevention, monitoring and retrieval device. Police can simply follow the signal emitted by the tracking system and locate the stolen vehicle. When used as a security system, a Vehicle Tracking System may serve as either an addition to or replacement for a traditional car alarm. Some vehicle tracking systems make it possible to control vehicle remotely, including block doors or engine in case of emergency. The existence of vehicle tracking device then can be used to reduce the insurance cost, because the loss-risk of the vehicle drops significantly.

Vehicle tracking systems are an integrated part of the "layered approach" to vehicle protection, recommended by the National Insurance Crime Bureau (NICB) to prevent motor vehicle theft. This approach recommends four layers of security based on the risk factors pertaining to a specific vehicle. Vehicle Tracking Systems are one such layer, and are described by the NICB as "very effective" in helping police recover stolen vehicles.

Some vehicle tracking systems integrate several security systems, for example by sending an automatic alert to a phone or email if an alarm is triggered or the vehicle is moved without authorization, or when it leaves or enters a geofence.

Other scenarios in which this technology is employed include:

- Stolen vehicle recovery: Both consumer and commercial vehicles can be outfitted with RF or GPS units to allow police to do tracking and recovery. In the case of LoJack, the police can activate the tracking unit in the vehicle directly and follow tracking signals.

- Fleet management: When managing a fleet of vehicles, knowing the real-time location of all drivers allows management to meet customer needs more efficiently. Whether it is delivery, service or other multi-vehicle enterprises, drivers now only need a mobile phone with telephony or Internet connection to be inexpensively tracked by and dispatched efficiently.

- Asset tracking: Companies needing to track valuable assets for insurance or other monitoring purposes can now plot the real-time asset location on a map and closely monitor movement and operating status.

- Field service management: Companies with a field service workforce for services such as repair or maintenance, must be able to plan field workers' time, schedule subsequent customer visits and be able to operate these departments efficiently. Vehicle tracking allows companies to quickly locate a field engineer and dispatch the closest one to meet a new customer request or provide site arrival information.

- Field sales: Mobile sales professionals can access real-time locations. For example, in unfamiliar areas, they can locate themselves as well as customers and prospects, get driving directions and add nearby last-minute appointments to itineraries. Benefits include increased productivity, reduced driving time and increased time spent with customers and prospects.

- Trailer tracking: Haulage and Logistics companies often operate lorries with detachable load carrying units. The part of the vehicle that drives the load is known as the cab and the load carrying unit is known as the trailer. There are different types of trailer used for different applications, e.g., flat bed, refrigerated, curtain sider, box container.

- Surveillance: A tracker may be placed on a vehicle to follow the vehicle's movements.

- Transit tracking: temporary tracking of assets or cargoes from one point to another. Users will ensure that the assets do not stop on route or do a U-Turn in order to ensure the security of the assets.

- Fuel Monitoring: monitor the fuel through tracking device (with help of fuel sensor connected to the device).

- Distance Calculation: calculate the distance traveled by the fleet.

- OBD II - Plug and play interface which provides most of engine diagnostics information.

Vehicle tracking systems are widely used worldwide. Components come in various shapes and forms but most utilize GPS technology and GSM services. While most will offer real-time tracking, others record real time data and store it to be read, in a fashion similar to data loggers. Systems like these track and record and allow reports after certain points have been solved.

## Unconventional Uses

Industries not traditionally known to use vehicle tracking systems (logistics and transportation industries are the ones that have traditionally incorporated vehicle tracking system into their operations) have started to use it in creative ways to improve their processes or businesses.

The hospitality industry have caught on to this technology to improve customer service. For example, a luxury hotel in Singapore has installed vehicle tracking systems in their limousines to ensure they can welcome their VIPs when they reach the hotel.

Vehicle tracking systems used in food delivery vans may alert if the temperature of the refrigerated compartment moves outside of the range of safe food storage temperatures. Car rental companies are also using it to monitor their rental fleets.

## Automatic Vehicle Location

Automatic vehicle location (AVL or ~locating; telelocating in EU) is a means for automatically determining and transmitting the geographic location of a vehicle. This data, from one or more vehicles, may then be collected by a vehicle tracking system for a picture of vehicle travel.

Most commonly, the location is determined using GPS, and the transmission mechanism is SMS, GPRS, a satellite or terrestrial radio from the vehicle to a radio receiver. GSM and EVDO are the most common services applied, because of the low data rate needed for AVL, and the low cost and near-ubiquitous nature of these public networks. The low bandwidth requirements also allow for satellite technology to receive telemetry data at a moderately higher cost, but across a global coverage area and into very remote locations not covered well by terrestrial radio or public carriers. Other options for determining actual location, for example in environments where GPS illumination is poor, are dead reckoning, i.e. inertial navigation, or active RFID systems or cooperative RTLS systems. With advantage, combinations of these systems may be applied. In addition, terrestrial radio positioning systems utilizing an LF (Low Frequency) switched packet radio network were also used as an alternative to GPS based systems.

# Applications

## Application with Vehicles

Automatic vehicle locating is a powerful concept for managing fleets of vehicles, as service vehicles, emergency vehicles, and especially precious construction equipment, also public transport vehicles (buses and trains). It is also used to track mobile assets, such as non wheeled construction equipment, non motorized trailers, and mobile power generators.

## Application with Vehicle Drivers and Crews

The other purpose of tracking is to provide graded service or to manage a large driver and crewing staff effectively. For example, suppose an ambulance fleet has an objective of arriving at the location of a call for service within six minutes of receiving the request. Using an AVL system allows to evaluate the locations of all vehicles in service with driver and other crew in order to pick the vehicle that will most likely arrive at the destination fastest, (meeting the service objective).

## Types of Systems

## Simple Direction Finding

Amateur radio and some cellular or PCS wireless systems use direction finding or triangulation of transmitter signals radiated by the mobile. This is sometimes called radio direction finding or RDF. The simplest forms of these systems calculate the bearing from two fixed sites to the mobile. This creates a triangle with endpoints at the two fixed points and the mobile. Trigonometry tells you roughly where the mobile transmitter is located. In wireless telephone systems, the phones transmit continually when off-hook, making continual tracking and the collection of many location samples possible. This is one type of location system required by Federal Communications Commission Rules for wireless Enhanced 911.

## Former LORAN-based Locating

Motorola offered a 1970s-era system based on the United States Coast Guard LORAN maritime navigation system. The LORAN system was intended for ships but signal levels on the US east- and west-coast areas were adequate for use with receivers in automobiles. The system may have been marketed under the Motorola model name *Metricom*. It consisted of an LF LORAN receiver and data interface box/modem connected to a separate two-way radio. The receiver and interface calculated a latitude and longitude in degrees, decimal degrees format based on the LORAN signals. This was sent over the radio as MDC-1200 or MDC-4800 data to a system controller, which plotted the mobile's approximate location on a map. The system worked reliably but sometimes had problems with electrical noise in urban areas. Sparking electric trolley poles or industrial plants which radiated electrical noise sometime overwhelmed the LORAN signals, affecting the system's ability to determine the mobile's geolocation. Because of the limited resolution, this type of system was impractical for small communities or operational areas such as a pit mine or port.

## Signpost Systems

To track and locate vehicles along fixed routes, a technology called Signpost transmitters is employed. This is used on transit routes and rail lines where the vehicles to be tracked continually operated on the same linear route. A transponder or RFID chip along the vehicle route would be polled as the train or bus traverses its route. As each transponder was passed, the moving vehicle would query and receive an ack, or handshake, from the signpost transmitter. A transmitter on the mobile would report passing the signpost to a system controller. This allows supervision, a call center, or a dispatch center to monitor the progress of the vehicle and assess whether or not the vehicle was *on schedule*. These systems are an alternative inside tunnels or other conveyances where GPS signals are blocked by terrain.

## Today's GPS-based Locating

The low price and ubiquity of Global Positioning System or GPS equipment has lent itself to more accurate and reliable telelocation systems. GPS signals are impervious to most electrical noise sources and don't require the user to install an entire system. Usually only a receiver to collect signals from the satellite segment is installed in each vehicle and radio or GSM to communicate the collected location data with a dispatch point.

Large private telelocation or AVL systems send data from GPS receivers in vehicles to a dispatch center over their private, user-owned radio backbone. These systems are used for businesses like parcel delivery and ambulances. Smaller systems which don't justify building a separate radio system use cellular or PCS data services to communicate location data from vehicles to their dispatching center. Location data is periodically polled from each vehicle in a fleet by a central controller or computer. In the simplest systems, data from the GPS receiver is displayed on a map allowing humans to determine the location of each vehicle. More complex systems feed the data into a computer assisted dispatch system which automates the process. For example, the computer assisted dispatch system may check the location of a call for service and then pick a list of the four closest ambulances. This narrows the dispatcher's choice from the entire fleet to an easier choice of four vehicles.

Some wireless carriers such as Nextel have decided GPS was the best way to provide the mandated location data for wireless Enhanced 9-1-1. Newer Nextel radios have embedded GPS receivers which are polled if 9-1-1 is dialed. The 9-1-1 center is provided with latitude and longitude from the radio's GPS receiver. In centers with computer assisted dispatch, the system may assign an address to the call based on these coordinates or may project an icon depicting the caller's location onto a map of the area.

## Sensor-augmented AVL

The main purpose of using AVL is not only to locate the vehicles, but also to obtain information about engine data, fuel consumption, driver data and sensor data from i.e. doors, freezer room on trucks or air pressure. Such data can be obtained via the CAN-bus, via direct connections to AVL systems or via open bus systems such as UFDEX that both sends and receives data via SMS or GPRS in pure ASCII text format. Because most AVL consists of two parts, GPS and GSM modem with additional embedded AVL software contained in a microcontroller, most AVL systems are fixed for its purposes unless they connect to an open bus system for expansion possibilities.

With an open bus system the users can send invoices based on goods delivered with exact location, time and date data where if connected to scale, RFID or barcode readers, can make a fairly good automated system to avoid human errors.

In countries with high prices on gasoline external fuel sensors are used to prevent cases of fuel theft.

## Logbook Functions

Another scenario for sensor functions is to connect the AVL to driver information, to collect data about driving time, stops, or even driver absence from the vehicle. If the driver/worker conditions is such as the hourly rates for driving and working outside is not the same, this can be monitored by sensors, by using iButton or other personal identification devices. Later by analyzing log-file it is possible to get reports on any kind of events, like stops, visited streets, speed limits violations, etc.

## Differentiating Between Automatic Vehicle Location and Events Activated Tracking Systems

It might be helpful to draw a distinction between vehicle location systems which track automatically and event activated tracking systems which track when triggered by an event. There is increasingly crossover between the different systems and those with experience of this sector will be able to draw on a number of examples which break the rule.

A.V.L (Automatic Vehicle Location) This type of vehicle tracking is normally used in the fleet or driver management sector. The unit is configured to automatically transmit its location at a set time interval, e.g. every 5 minutes. The unit is activated when the ignition is switched on/off.

E.A.T.S (Events Activated Tracking system) This type of system is primarily used in connection with vehicle or driver security solutions. If, for example a thief breaks into your car and attempts to steal it, the tracking system can be triggered by the immobiliser unit or motion sensor being activated. A monitoring bureau, will then be automatically notified that the unit has been activated and begin tracking the vehicle.

Some products on the market are a hybrid of both AVL and EATS technology. However industry practice has tended to lean towards a separation of these functions. It is worth taking note that vehicle tracking products tend to fall into one, not both of the technologies.

AVL technology is predominately used when applying vehicle tracking to fleet or driver management solutions. The use of Automatic Vehicle Location is given in the following scenario; A car breaks down by the side of the road and the occupant calls a vehicle recovery company. The vehicle recovery company has several vehicles operating in the area. Without needing to call each driver to check his location the dispatcher can pinpoint the nearest recovery vehicle and assign it to the new job. If you were to incorporate the other aspects of vehicle telematics into this scenario; the dispatcher, rather than phoning the recovery vehicle operative, could transmit the job details directly to the operative's mobile data device, who would then use the in-vehicle satellite navigation to aid his journey to the job.

EATS technology is predominately used when applying vehicle tracking to vehicle security solutions. An example of this distinction is given in the following scenario; A construction company owns some pieces of plant machinery that are regularly left unattended, at weekends, on building sites. Thieves break onto one site and a piece equipment, such as a digger, is loaded on the back of a flat bed truck and then driven away. Typically the ignition wouldn't need to be turned on and as such most of the AVL products available wouldn't typically be activated. Only products that included a unit that was activated by a motion sensor or GeoFence alarm event, would be activated.

Both AVL and EATS systems track, but often for different purposes.

## Special Applications of Automatic Vehicle Locating

Vehicle location technologies can be used in the following scenarios:

- Fleet management: when managing a fleet of vehicles, knowing the real-time location of all drivers allows management to meet customer needs more efficiently. Vehicle location information can also be used to verify that legal requirements are being met: for example, that drivers are taking rest breaks and obeying speed limits.

- Passenger Information: Real-time Passenger information systems use predictions based on AVL input to show the expected arrival and departure times of Public Transport services.

- Asset tracking: companies needing to track valuable assets for insurance or other monitoring purposes can now plot the real-time asset location on a map and closely monitor movement and operating status. For example, haulage and logistics companies often operate trucks with detachable load carrying units. In this case, trailers can be tracked independently of the cabs used to drive them. Combining vehicle location with inventory management that can be used to reconcile which item is currently on which vehicle can be used to identify physical location down to the level of individual packages.

- Field worker management: companies with a field service or sales workforce can use information from vehicle tracking systems to plan field workers' time, schedule subsequent customer visits and be able to operate these departments efficiently.

- Covert surveillance: vehicle location devices attached covertly by law enforcement or espionage organizations can be used to track journeys made by individuals who are under surveillance

# Satellite: An Overview

All navigation systems are dependent on satellites orbiting the atmosphere for providing accurate information. A satellite is a man-made object that is launched into the orbit of the Earth by a spacecraft. This chapter illustrates the major satellite subsystems such as satellite bus, fractioned spacecraft, telemetry, GNSS software-defined receiver and spacecraft propulsion. The chapter serves as a source to understand the major categories related to satellites.

## Satellite

A full-size model of the Earth observation satellite ERS 2

In the context of spaceflight, a satellite is an artificial object which has been intentionally placed into orbit. Such objects are sometimes called artificial satellites to distinguish them from natural satellites such as Earth's Moon.

The world's first artificial satellite, the Sputnik 1, was launched by the Soviet Union in 1957. Since then, thousands of satellites have been launched into orbit around the Earth. Some satellites, notably space stations, have been launched in parts and assembled in orbit. Artificial satellites originate from more than 40 countries and have used the satellite launching capabilities of ten nations. About a thousand satellites are currently operational, whereas thousands of unused satellites and satellite fragments orbit the Earth as space debris. A few space probes have been placed into orbit around other bodies and become artificial satellites to the Moon, Mercury, Venus, Mars, Jupiter, Saturn, Vesta, Eros, Ceres, and the Sun.

Satellites are used for a large number of purposes. Common types include military and civilian Earth observation satellites, communications satellites, navigation satellites, weather satellites, and research satellites. Space stations and human spacecraft in orbit are also satellites. Satellite orbits vary greatly, depending on the purpose of the satellite, and are classified in a number of ways. Well-known (overlapping) classes include low Earth orbit, polar orbit, and geostationary orbit.

About 6,600 satellites have been launched. The latest estimates are that 3,600 remain in orbit. Of those, about 1,000 are operational; the rest have lived out their useful lives and are part of the space debris. Approximately 500 operational satellites are in low-Earth orbit, 50 are in medium-Earth orbit (at 20,000 km), and the rest are in geostationary orbit (at 36,000 km).

Satellites are propelled by rockets to their orbits. Usually the launch vehicle itself is a rocket lifting off from a launch pad on land. In a minority of cases satellites are launched at sea (from a submarine or a mobile maritime platform) or aboard a plane.

Satellites are usually semi-independent computer-controlled systems. Satellite subsystems attend many tasks, such as power generation, thermal control, telemetry, attitude control and orbit control.

## History

### Early Conceptions

"Newton's cannonball", presented as a "thought experiment" in *A Treatise of the System of the World*, by Isaac Newton was the first published mathematical study of the possibility of an artificial satellite.

The first fictional depiction of a satellite being launched into orbit is a short story by Edward Everett Hale, *The Brick Moon*. The story is serialized in *The Atlantic Monthly*, starting in 1869. The idea surfaces again in Jules Verne's *The Begum's Fortune* (1879).

Konstantin Tsiolkovsky

In 1903, Konstantin Tsiolkovsky (1857–1935) published *Exploring Space Using Jet Propulsion Devices* (in Russian: *Исследование мировых пространств реактивными приборами*), which is the first academic treatise on the use of rocketry to launch spacecraft. He calculated the orbital speed required for a minimal orbit around the Earth at 8 km/s, and that a multi-stage rocket fuelled by liquid propellants could be used to achieve this. He proposed the use of liquid hydrogen and liquid oxygen, though other combinations can be used.

In 1928, Slovenian Herman Potočnik (1892–1929) published his sole book, *The Problem of Space Travel — The Rocket Motor* (German: *Das Problem der Befahrung des Weltraums — der Raketen-Motor*), a plan for a breakthrough into space and a permanent human presence there. He conceived of a space station in detail and calculated its geostationary orbit. He described the use of orbiting spacecraft for detailed peaceful and military observation of the ground and described how the special conditions of space could be useful for scientific experiments. The book described geostationary satellites (first put forward by Tsiolkovsky) and discussed communication between them and the ground using radio, but fell short of the idea of using satellites for mass broadcasting and as telecommunications relays.

In a 1945 *Wireless World* article, the English science fiction writer Arthur C. Clarke (1917–2008) described in detail the possible use of communications satellites for mass communications. Clarke examined the logistics of satellite launch, possible orbits and other aspects of the creation of a network of world-circling satellites, pointing to the benefits of high-speed global communications. He also suggested that three geostationary satellites would provide coverage over the entire planet.

The US military studied the idea of what was referred to as the *earth satellite vehicle* when Secretary of Defense James Forrestal made a public announcement on December 29, 1948, that his office was coordinating that project between the various services.

## Artificial Satellites

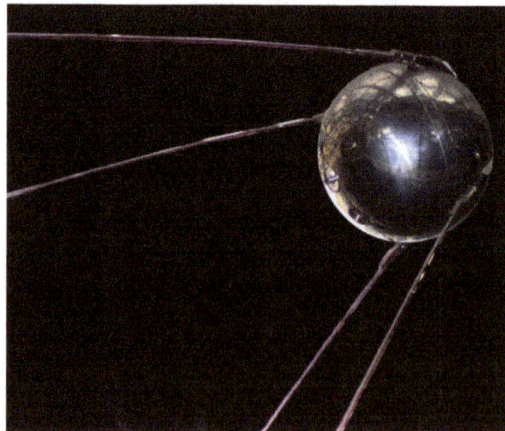

Sputnik 1: The first artificial satellite to orbit Earth.

The first artificial satellite was Sputnik 1, launched by the Soviet Union on October 4, 1957, and initiating the Soviet Sputnik program, with Sergei Korolev as chief designer (there is a crater on the lunar far side which bears his name). This in turn triggered the Space Race between the Soviet Union and the United States.

Sputnik 1 helped to identify the density of high atmospheric layers through measurement of its orbital change and provided data on radio-signal distribution in the ionosphere. The unanticipated announcement of *Sputnik 1*'s success precipitated the Sputnik crisis in the United States and ignited the so-called Space Race within the Cold War.

Sputnik 2 was launched on November 3, 1957 and carried the first living passenger into orbit, a dog named Laika.

In May, 1946, Project RAND had released the Preliminary Design of an Experimental World-Circling Spaceship, which stated, "A satellite vehicle with appropriate instrumentation can be expected to be one of the most potent scientific tools of the Twentieth Century." The United States had been considering launching orbital satellites since 1945 under the Bureau of Aeronautics of the United States Navy. The United States Air Force's Project RAND eventually released the above report, but did not believe that the satellite was a potential military weapon; rather, they considered it to be a tool for science, politics, and propaganda. In 1954, the Secretary of Defense stated, "I know of no American satellite program." In February 1954 Project RAND released "Scientific Uses for a Satellite Vehicle," written by R.R. Carhart. This expanded on potential scientific uses for satellite vehicles and was followed in June 1955 with "The Scientific Use of an Artificial Satellite," by H.K. Kallmann and W.W. Kellogg.

In the context of activities planned for the International Geophysical Year (1957–58), the White House announced on July 29, 1955 that the U.S. intended to launch satellites by the spring of 1958. This became known as Project Vanguard. On July 31, the Soviets announced that they intended to launch a satellite by the fall of 1957.

Following pressure by the American Rocket Society, the National Science Foundation, and the International Geophysical Year, military interest picked up and in early 1955 the Army and Navy were working on Project Orbiter, two competing programs: the army's which involved using a Jupiter C rocket, and the civilian/Navy Vanguard Rocket, to launch a satellite. At first, they failed: initial preference was given to the Vanguard program, whose first attempt at orbiting a satellite resulted in the explosion of the launch vehicle on national television. But finally, three months after Sputnik 2, the project succeeded; Explorer 1 became the United States' first artificial satellite on January 31, 1958.

In June 1961, three-and-a-half years after the launch of Sputnik 1, the Air Force used resources of the United States Space Surveillance Network to catalog 115 Earth-orbiting satellites.

Early satellites were constructed as "one-off" designs. With growth in geosynchronous (GEO) satellite communication, multiple satellites began to be built on single model platforms called satellite buses. The first standardized satellite bus design was the HS-333 GEO commsat, launched in 1972.

The largest artificial satellite currently orbiting the Earth is the International Space Station.

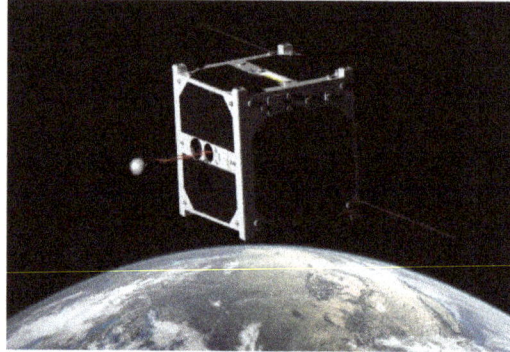

1U CubeSat ESTCube-1, developed mainly by the students from the University of Tartu, carries out a tether deployment experiment on the low Earth orbit.

## Space Surveillance Network

The United States Space Surveillance Network (SSN), a division of the United States Strategic Command, has been tracking objects in Earth's orbit since 1957 when the Soviet Union opened the Space Age with the launch of Sputnik I. Since then, the SSN has tracked more than 26,000 objects. The SSN currently tracks more than 8,000 man-made orbiting objects. The rest have re-entered Earth's atmosphere and disintegrated, or survived re-entry and impacted the Earth. The SSN tracks objects that are 10 centimeters in diameter or larger; those now orbiting Earth range from satellites weighing several tons to pieces of spent rocket bodies weighing only 10 pounds. About seven percent are operational satellites (i.e. ~560 satellites), the rest are space debris. The United States Strategic Command is primarily interested in the active satellites, but also tracks space debris which upon reentry might otherwise be mistaken for incoming missiles.

A search of the NSSDC Master Catalog at the end of October 2010 listed 6,578 satellites launched into orbit since 1957, the latest being Chang'e 2, on 1 October 2010.

## Non-military Satellite Services

There are three basic categories of non-military satellite services:

## Fixed Satellite Services

Fixed satellite services handle hundreds of billions of voice, data, and video transmission tasks across all countries and continents between certain points on the Earth's surface.

## Mobile Satellite Systems

Mobile satellite systems help connect remote regions, vehicles, ships, people and air-craft to other parts of the world and/or other mobile or stationary communications units, in addition to serving as navigation systems.

## Scientific Research Satellites (Commercial and Noncommercial)

Scientific research satellites provide meteorological information, land survey data (e.g. remote sensing), Amateur (HAM) Radio, and other different scientific research appli-cations such as earth science, marine science, and atmospheric research.

## Types

- Astronomical satellites are satellites used for observation of distant planets, gal-axies, and other outer space objects.

- Biosatellites are satellites designed to carry living organisms, generally for sci-entific experimentation.

- Communications satellites are satellites stationed in space for the purpose of telecommunications. Modern communications satellites typically use geosyn-chronous orbits, Molniya orbits or Low Earth orbits.

- Earth observation satellites are satellites intended for non-military uses such as environmental monitoring, meteorology, map making etc.

- Navigational satellites are satellites which use radio time signals transmit-ted to enable mobile receivers on the ground to determine their exact loca-tion. The relatively clear line of sight between the satellites and receivers on the ground, combined with ever-improving electronics, allows satellite navigation systems to measure location to accuracies on the order of a few meters in real time.

- "Killer Satellites" are satellites that are designed to destroy enemy warheads, satellites, and other space assets.

- Crewed spacecraft (spaceships) are large satellites able to put humans into (and beyond) an orbit, and return them to Earth. Spacecraft including spaceplanes of reusable systems have major propulsion or landing facilities. They can be used as transport to and from the orbital stations.

- Miniaturized satellites are satellites of unusually low masses and small siz-es. New classifications are used to categorize these satellites: minisatellite (500– 100 kg), microsatellite (below 100 kg), nanosatellite (below 10 kg).

- Reconnaissance satellites are Earth observation satellite or communications satellite deployed for military or intelligence applications. Very little is known about the full power of these satellites, as governments who operate them usually keep information pertaining to their reconnaissance satellites classified.

- Recovery satellites are satellites that provide a recovery of reconnaissance, biological, space-production and other payloads from orbit to Earth.

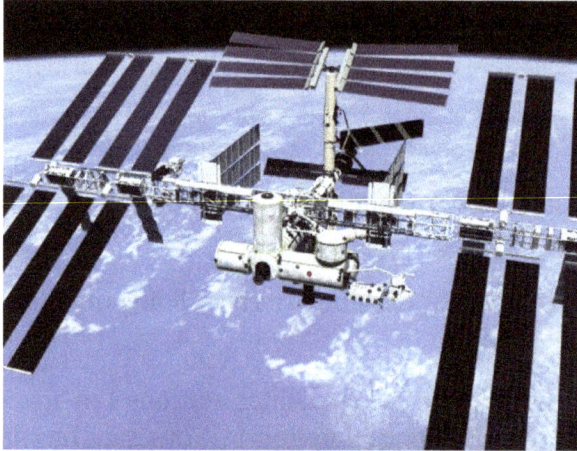

artist's depiction of the International Space Station

- Space stations are artificial orbital structures that are designed for human beings to live on in outer space. A space station is distinguished from other crewed spacecraft by its lack of major propulsion or landing facilities. Space stations are designed for medium-term living in orbit, for periods of weeks, months, or even years.

- Tether satellites are satellites which are connected to another satellite by a thin cable called a tether.

- Weather satellites are primarily used to monitor Earth's weather and climate.

## Orbit Types

The first satellite, Sputnik 1, was put into orbit around Earth and was therefore in geocentric orbit. By far this is the most common type of orbit with approximately 2,465 artificial satellites orbiting the Earth. Geocentric orbits may be further classified by their altitude, inclination and eccentricity.

The commonly used altitude classifications of geocentric orbit are Low Earth orbit (LEO), Medium Earth orbit (MEO) and High Earth orbit (HEO). Low Earth orbit is any orbit below 2,000 km. Medium Earth orbit is any orbit between 2,000 and 35,786 km. High Earth orbit is any orbit higher than 35,786 km.

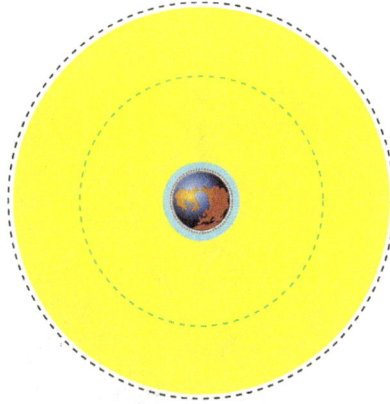

Various earth orbits to scale; cyan represents low earth orbit, yellow represents medium earth orbit, the black dashed line represents geosynchronous orbit, the green dash-dot line the orbit of Global Positioning System (GPS) satellites, and the red dotted line the orbit of the International Space Station (ISS).

## Centric Classifications

- Geocentric orbit: An orbit around the planet Earth, such as the Moon or artificial satellites. Currently there are approximately 1,071 artificial satellites orbiting the Earth.

- Heliocentric orbit: An orbit around the Sun. In our Solar System, all planets, comets, and asteroids are in such orbits, as are many artificial satellites and pieces of space debris. Moons by contrast are not in a heliocentric orbit but rather orbit their parent planet.

- Areocentric orbit: An orbit around the planet Mars, such as by moons or artificial satellites.

The general structure of a satellite is that it is connected to the earth stations that are present on the ground and connected through terrestrial links.

## Altitude Classifications

- Low Earth orbit (LEO): Geocentric orbits ranging in altitude from -0.428 km - 2,000 km (1,200 mi)

- Medium Earth orbit (MEO): Geocentric orbits ranging in altitude from 2,000 km (1,200 mi) - 35,786 km (22,236 mi). Also known as an intermediate circular orbit.

- Geosynchronous Orbit (GEO): Geocentric circular orbit with an altitude of 35,786 kilometres (22,236 mi). The period of the orbit equals one sidereal day, coinciding with the rotation period of the Earth. The speed is approximately 3,000 metres per second (9,800 ft/s).

- High Earth orbit (HEO): Geocentric orbits above the altitude of geosynchronous orbit 35,786 km (22,236 mi).

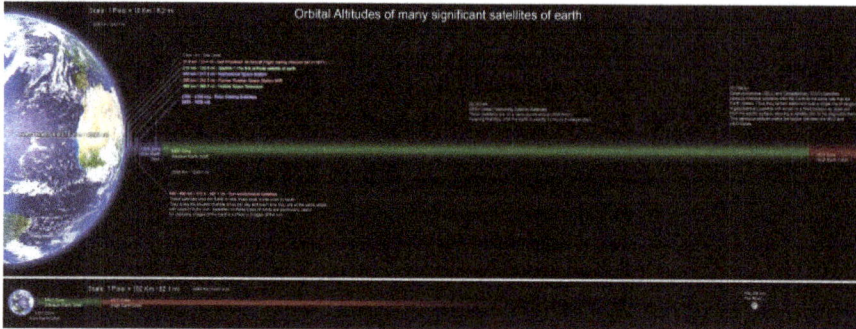

Orbital Altitudes of several significant satellites of earth.

## Inclination Classifications

- Inclined orbit: An orbit whose inclination in reference to the equatorial plane is not zero degrees.

  o Polar orbit: An orbit that passes above or nearly above both poles of the planet on each revolution. Therefore, it has an inclination of (or very close to) 90 degrees.

  o Polar sun synchronous orbit: A nearly polar orbit that passes the equator at the same local time on every pass. Useful for image taking satellites because shadows will be nearly the same on every pass.

Eccentricity classifications

- Circular orbit: An orbit that has an eccentricity of 0 and whose path traces a circle.

  o Hohmann transfer orbit: An orbit that moves a spacecraft from one approximately circular orbit, usually the orbit of a planet, to another, using two engine impulses. The perihelion of the transfer orbit is at the same distance from the Sun as the radius of one planet's orbit, and the aphelion is at the other. The two rocket burns change the spacecraft's path from one circular orbit to the transfer orbit, and later to the other circular orbit. This maneuver was named after Walter Hohmann.

- Elliptic orbit: An orbit with an eccentricity greater than 0 and less than 1 whose orbit traces the path of an ellipse.

  o Geosynchronous transfer orbit: An elliptic orbit where the perigee is at the altitude of a Low Earth orbit (LEO) and the apogee at the altitude of a geosynchronous orbit.

- o  Geostationary transfer orbit: An elliptic orbit where the perigee is at the altitude of a Low Earth orbit (LEO) and the apogee at the altitude of a geostationary orbit.

- o  Molniya orbit: A highly elliptic orbit with inclination of 63.4° and orbital period of half of a sidereal day (roughly 12 hours). Such a satellite spends most of its time over two designated areas of the planet (specifically Russia and the United States).

- o  Tundra orbit: A highly elliptic orbit with inclination of 63.4° and orbital period of one sidereal day (roughly 24 hours). Such a satellite spends most of its time over a single designated area of the planet.

## Synchronous Classifications

- •  Synchronous orbit: An orbit where the satellite has an orbital period equal to the average rotational period (earth's is: 23 hours, 56 minutes, 4.091 seconds) of the body being orbited and in the same direction of rotation as that body. To a ground observer such a satellite would trace an analemma (figure 8) in the sky.

- •  Semi-synchronous orbit (SSO): An orbit with an altitude of approximately 20,200 km (12,600 mi) and an orbital period equal to one-half of the average rotational period (Earth's is approximately 12 hours) of the body being orbited

- •  Geosynchronous orbit (GSO): Orbits with an altitude of approximately 35,786 km (22,236 mi). Such a satellite would trace an analemma (figure 8) in the sky.

  - o  Geostationary orbit (GEO): A geosynchronous orbit with an inclination of zero. To an observer on the ground this satellite would appear as a fixed point in the sky.

    - ▪  Clarke orbit: Another name for a geostationary orbit. Named after scientist and writer Arthur C. Clarke.

  - o  Supersynchronous orbit: A disposal / storage orbit above GSO/GEO. Satellites will drift west. Also a synonym for Disposal orbit.

  - o  Subsynchronous orbit: A drift orbit close to but below GSO/GEO. Satellites will drift east.

  - o  Graveyard orbit: An orbit a few hundred kilometers above geosynchronous that satellites are moved into at the end of their operation.

    - ▪  Disposal orbit: A synonym for graveyard orbit.

    - ▪  Junk orbit: A synonym for graveyard orbit.

- Areosynchronous orbit: A synchronous orbit around the planet Mars with an orbital period equal in length to Mars' sidereal day, 24.6229 hours.

- Areostationary orbit (ASO): A circular areosynchronous orbit on the equatorial plane and about 17000 km (10557 miles) above the surface. To an observer on the ground this satellite would appear as a fixed point in the sky.

- Heliosynchronous orbit: A heliocentric orbit about the Sun where the satellite's orbital period matches the Sun's period of rotation. These orbits occur at a radius of 24,360 Gm (0.1628 AU) around the Sun, a little less than half of the orbital radius of Mercury.

## Special Classifications

- Sun-synchronous orbit: An orbit which combines altitude and inclination in such a way that the satellite passes over any given point of the planets' surface at the same local solar time. Such an orbit can place a satellite in constant sunlight and is useful for imaging, spy, and weather satellites.

- Moon orbit: The orbital characteristics of Earth's Moon. Average altitude of 384,403 kilometers (238,857 mi), elliptical–inclined orbit.

## Pseudo-orbit Classifications

- Horseshoe orbit: An orbit that appears to a ground observer to be orbiting a certain planet but is actually in co-orbit with the planet.

- Exo-orbit: A maneuver where a spacecraft approaches the height of orbit but lacks the velocity to sustain it.

    o Suborbital spaceflight: A synonym for exo-orbit.

- Lunar transfer orbit (LTO)

- Prograde orbit: An orbit with an inclination of less than 90°. Or rather, an orbit that is in the same direction as the rotation of the primary.

- Retrograde orbit: An orbit with an inclination of more than 90°. Or rather, an orbit counter to the direction of rotation of the planet. Apart from those in sun-synchronous orbit, few satellites are launched into retrograde orbit because the quantity of fuel required to launch them is much greater than for a prograde orbit. This is because when the rocket starts out on the ground, it already has an eastward component of velocity equal to the rotational velocity of the planet at its launch latitude.

- Halo orbit and Lissajous orbit: Orbits "around" Lagrangian points.

## Satellite Subsystems

The satellite's functional versatility is imbedded within its technical components and its operations characteristics. Looking at the "anatomy" of a typical satellite, one discovers two modules. Note that some novel architectural concepts such as Fractionated spacecraft somewhat upset this taxonomy.

## Spacecraft Bus or Service Module

The bus module consists of the following subsystems:

## Structural Subsystem

The structural subsystem provides the mechanical base structure with adequate stiffness to withstand stress and vibrations experienced during launch, maintain structural integrity and stability while on station in orbit, and shields the satellite from extreme temperature changes and micro-meteorite damage.

## Telemetry Subsystem

The telemetry subsystem (aka Command and Data Handling, C&DH) monitors the onboard equipment operations, transmits equipment operation data to the earth control station, and receives the earth control station's commands to perform equipment operation adjustments.

## Power Subsystem

The power subsystem consists of solar panels to convert solar energy into electrical power, regulation and distribution functions, and batteries that store power and supply the satellite when it passes into the Earth's shadow. Nuclear power sources (Radioisotope thermoelectric generator have also been used in several successful satellite programs including the Nimbus program (1964–1978).

## Thermal Control Subsystem

The thermal control subsystem helps protect electronic equipment from extreme temperatures due to intense sunlight or the lack of sun exposure on different sides of the satellite's body (e.g. Optical Solar Reflector)

## Attitude and Orbit Control Subsystem

The attitude and orbit control subsystem consists of sensors to measure vehicle orientation; control laws embedded in the flight software; and actuators (reaction wheels, thrusters) to apply the torques and forces needed to re-orient the vehicle to a desired attitude, keep the satellite in the correct orbital position and keep antennas positioning in the right directions.

## Communication Payload

The second major module is the communication payload, which is made up of transponders. A transponder is capable of :

- Receiving uplinked radio signals from earth satellite transmission stations (antennas).

- Amplifying received radio signals

- Sorting the input signals and directing the output signals through input/output signal multiplexers to the proper downlink antennas for retransmission to earth satellite receiving stations (antennas).

## End of Life

When satellites reach the end of their mission, satellite operators have the option of de-orbiting the satellite, leaving the satellite in its current orbit or moving the satellite to a graveyard orbit. Historically, due to budgetary constraints at the beginning of satellite missions, satellites were rarely designed to be de-orbited. One example of this practice is the satellite Vanguard 1. Launched in 1958, Vanguard 1, the 4th manmade satellite put in Geocentric orbit, was still in orbit as of August 2009.

Instead of being de-orbited, most satellites are either left in their current orbit or moved to a graveyard orbit. As of 2002, the FCC requires all geostationary satellites to commit to moving to a graveyard orbit at the end of their operational life prior to launch. In cases of uncontrolled de-orbiting, the major variable is the solar flux, and the minor variables the components and form factors of the satellite itself, and the gravitational perturbations generated by the Sun and the Moon (as well as those exercised by large mountain ranges, whether above or below sea level). The nominal breakup altitude due to aerodynamic forces and temperatures is 78 km, with a range between 72 and 84 km. Solar panels, however, are destroyed before any other component at altitudes between 90 and 95 km.

## Launch-capable Countries

This list includes countries with an independent capability of states to place satellites in orbit, including production of the necessary launch vehicle. Note: many more countries have the capability to design and build satellites but are unable to launch them, instead relying on foreign launch services. This list does not consider those numerous countries, but only lists those capable of launching satellites indigenously, and the date this capability was first demonstrated. The list includes the European Space Agency, a multi-national state organization, but does not include private consortiums.

| First launch by country | | | | |
|---|---|---|---|---|
| **Order** | **Country** | **Date of first launch** | **Rocket** | **Satellite** |
| 1 | Soviet Union | 4 October 1957 | Sputnik-PS | Sputnik 1 |
| 2 | United States | 1 February 1958 | Juno I | Explorer 1 |
| 3 | France | 26 November 1965 | Diamant-A | Astérix |
| 4 | Japan | 11 February 1970 | Lambda-4S | Ōsumi |
| 5 | China | 24 April 1970 | Long March 1 | Dong Fang Hong I |
| 6 | United Kingdom | 28 October 1971 | Black Arrow | Prospero |
| 7 | India | 18 July 1980 | SLV | Rohini D1 |
| 8 | Israel | 19 September 1988 | Shavit | Ofeq 1 |
| - | Russia | 21 January 1992 | Soyuz-U | Kosmos 2175 |
| - | Ukraine | 13 July 1992 | Tsyklon-3 | Strela |
| 9 | Iran | 2 February 2009 | Safir-1 | Omid |
| 10 | North Korea | 12 December 2012 | Unha-3 | Kwangmyŏngsŏng-3 Unit 2 |

## Attempted First Launches

- The United States tried in 1957 to launch the first satellite using its own launcher before successfully completing a launch in 1958.

- China tried in 1969 to launch the first satellite using its own launcher before successfully completing a launch in 1970.

- India, after launching its first national satellite using a foreign launcher in 1975, tried in 1979 to launch the first satellite using its own launcher before succeeding in 1980.

- Iraq have claimed an orbital launch of a warhead in 1989, but this claim was later disproved.

- Brazil, after launching its first national satellite using a foreign launcher in 1985, tried to launch a satellite using its own VLS 1 launcher three times in 1997, 1999, and 2003, but all attempts were unsuccessful.

- North Korea claimed a launch of Kwangmyŏngsŏng-1 and Kwangmyŏngsŏng-2 satellites in 1998 and 2009, but U.S., Russian and other officials and weapons experts later reported that the rockets failed to send a satellite into orbit, if that was the goal. The United States, Japan and South Korea believe this was actually a ballistic missile test, which was a claim also made after North Korea's 1998 satellite launch, and later rejected.[by whom?] The first (April 2012) launch of Kwangmyŏngsŏng-3 was unsuccessful, a fact publicly recognized by the DPRK. However, the December 2012 launch of the "second version" of Kwangmyŏngsŏng-3 was successful, putting the DPRK's first confirmed satellite into orbit.

- South Korea (Korea Aerospace Research Institute), after launching their first national satellite by foreign launcher in 1992, unsuccessfully tried to launch its own launcher, the KSLV(Naro)-1, (created with the assistance of Russia) in 2009 and 2010 until success was achieved in 2013 by Naro-3.

- The First European multi-national state organization ELDO tried to make the orbital launches at Europa I and Europa II rockets in 1968-1970 and 1971 but stopped operation after failures.

## Other Notes

- Russia and the Ukraine were parts of the Soviet Union and thus inherited their launch capability without the need to develop it indigenously. Through the Soviet Union they are also on the number one position in this list of accomplishments.

- France, the United Kingdom, and Ukraine launched their first satellites by own launchers from foreign spaceports.

- Some countries such as South Africa, Spain, Italy, Germany, Canada, Australia, Argentina, Egypt and private companies such as OTRAG, have developed their own launchers, but have not had a successful launch.

- Only twelve, countries from the list below (USSR, USA, France, Japan, China, UK, India, Russia, Ukraine, Israel, Iran and North Korea) and one regional organization (the European Space Agency, ESA) have independently launched satellites on their own indigenously developed launch vehicles.

- Several other countries, including Brazil, Argentina, Pakistan, Romania, Taiwan, Indonesia, Australia, New Zealand, Malaysia, Turkey and Switzerland are at various stages of development of their own small-scale launcher capabilities.

## Launch Capable Private Entities

- Private firm Orbital Sciences Corporation, with launches since 1982, continues very successful launches of its Minotaur, Pegasus, Taurus and Antares rocket programs.

- On September 28, 2008, late comer and private aerospace firm SpaceX successfully launched its Falcon 1 rocket into orbit. This marked the first time that a privately built liquid-fueled booster was able to reach orbit. The rocket carried a prism shaped 1.5 m (5 ft) long payload mass simulator that was set into orbit. The dummy satellite, known as Ratsat, will remain in orbit for between five and ten years before burning up in the atmosphere.

A few other private companies are capable of sub-orbital launches.

## First Satellites of Countries

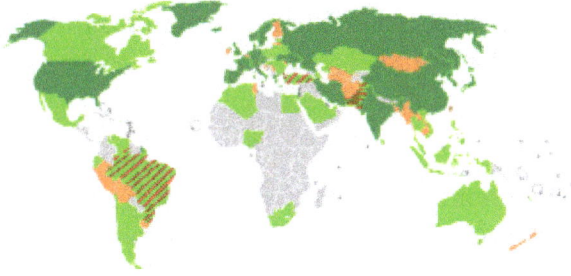

| orbital launch and satellite operation
| satellite operation, launched by foreign supplier
| satellite in development
| orbital launch project at advanced stage or indigenous ballistic missiles deployed

While Canada was the third country to build a satellite which was launched into space, it was launched aboard an American rocket from an American spaceport. The same goes for Australia, who launched first satellite involved a donated U.S. Redstone rocket and American support staff as well as a joint launch facility with the United Kingdom. The first Italian satellite San Marco 1 launched on 15 December 1964 on a U.S. Scout rocket from Wallops Island (Virginia, United States) with an Italian launch team trained by NASA. By similar occasions, almost all further first national satellites was launched by foreign rockets.

## Attempted First Satellites

- United States tried unsuccessfully to launch its first satellite in 1957; they were successful in 1958.

- China tried unsuccessfully to launch its first satellite in 1969; they were successful in 1970.

- Iraq under Saddam fulfilled in 1989 an unconfirmed launch of warhead on orbit by developed Iraqi vehicle that intended to put later the 75 kg first national satellite Al-Ta'ir, also developed.

- Chile tried unsuccessfully in 1995 to launch its first satellite *FASat-Alfa* by foreign rocket; in 1998 they were successful.[†]

- North Korea has tried in 1998, 2009, 2012 to launch satellites, first successful launch on 12 December 2012.

- Libya since 1996 developed its own national Libsat satellite project with the goal of providing telecommunication and remote sensing services that was postponed after the fall of Gaddafi.

- Belarus tried unsuccessfully in 2006 to launch its first satellite *BelKA* by foreign rocket.†

†-note: Both Chile and Belarus used Russian companies as principal contractors to build their satellites, they used Russian-Ukrainian manufactured rockets and launched either from Russia or Kazakhstan.

## Planned First Satellites

- Afghanistan announced in April 2012 that it is planning to launch its first communications satellite to the orbital slot it has been awarded. The satellite Afghansat 1 was expected to be obtained by a Eutelsat commercial company in 2014.

- Angola will have the first telecommunication satellite AngoSat-1 that was ordered in Russia at 2009 for $400 millions, started to construction at the end of 2013 and planning for launch in November 2016.

- Armenia in 2012 founded Armcosmos company and announced an intenton to have the first telecommunication satellite ArmSat. The investments estimates as $250 million and country selecting the contractor for building within 4 years the satellite amongst Russia, China and Canada

- Bangladesh announced in 2009 that it intends to launch its first satellite into space by 2011.

- Cambodia's Royal Group plans to purchase for $250–350 million and launch in the beginning of 2013 the telecommunication satellite.

- Democratic Republic of the Congo ordered at November 2012 in China (Academy of Space Technology (CAST) and Great Wall Industry Corporation (CGWIC)) the first telecommunication satellite CongoSat-1 which will be built on DFH-4 satellite bus platform and will be launched in China till the end of 2015.

- Croatia has a goal to construct a satellite by 2013–2014. Launch into Earth orbit would be done by a foreign provider.

- Ethiopian Space Science Society planning the QB50-family research CubeSat ET-SAT by help of Belgian Von Karman Institute till 2015 and the small (20–25 kg) Earth observation and remote sensing satellite Ethosat 1 by help of Finnish Space Technology and Science Group till 2019.

- ✚ Finland's Aalto-1 Cusesat-satellite (3U) with solar panels is a funded by student nano-satellite project of Aalto University and Finnish Meteorological Institute . When launched (plan was to 2013), it would be the first Finnish satellite. Launch has been procured for the summer 2015.

- Ghana plans to order in United Kingdom and Italy and launch within 2020 the first Earth observation satellite Ghanasat-1.

- Ireland's team of Dublin Institute of Technology intends to launch the first Irish satellite within European University program CubeSat QB50.

- Jordan's first satellite to be the private amateur pocketqube SunewnewSat.

- Kenyan University of Nairobi has plans to create the microsatellite Kenya-Sat by help of UK's University of Surrey.

- Latvia's the 5 kg nano-satellite Venta-1 is built in Latvia in cooperation with the German engineers. The data received from satellite will be received and processed in Irbene radioastronomical centre (Latvia); satellite will have software defined radio capabilities. "Venta-1" will serve mainly as a means for education in Ventspils University College with additional functions, including an automatic system of identification of the ships of a sailing charter developed by *OHB-System AG*. The launch of the satellite was planned for the end of 2009 using the Indian carrier rocket. Due to the financial crisis the launch has been postponed until late 2011. Started preparations to produce the next satellite "Venta-2".

- Moldova's first remote sensing satellite plans to start in 2013 by Space centre at national Technical University.

- Mongolia's National Remote Sensing Center of Mongolia plans to order the communication satellite in Japan, Mongolian Academy of Sciences schedules to launch the first national experimental satellite Mongolsat by US launcher in the first quarter of 2013.

- Myanmar plans to purchase for $200 million the own telecommunication satellite.

- Nepal stated that planning to launch of own telecommunication satellite before 2015 by help of India or China.

- New Zealand's private Satellite Opportunities company since 2005 plans to launch in 2010 or later a commercial satellite NZLSAT for $200 million. Radio enthusiasts federation at Massey University  since 2003 hopes for $400,000 to launch a nano-satellite KiwiSAT to relay a voice and data signals Also another RocketLab company works under suborbital space launcher and may use a further version of one to launch into low polar orbit a nano-satellite.

- Nicaragua ordered for $254 million at November 2013 in China the first telecommunication satellite Nicasat-1 (to be built at DFH-4 satellite bus platform by CAST and CGWIC), that planning to launch in China at 2016.

- Paraguay under new Aaepa airspace agency plans first Eart observation satellite.

- Serbia's first satellite Tesla-1 was designed, developed and assembled by nongovermental organisations in 2009 but still remains unlaunched.

- Slovakian Organisation for Space Activities (SOSA) together with University of Žilina and Slovak University of Technology developing the first national satellite SkCube under European University program CubeSat QB50 since 2012 aiming to launch them in 2016.

- Slovenia's Earth observation microsatellite for the Slovenian Centre of Excellence for Space Sciences and Technologies (Space-SI) now under development for $2 million since 2010 by University of Toronto Institute for Aerospace Studies – Space Flight Laboratory (UTIAS – SFL) and planned to launch in 2015-2016.

- Sri Lanka has a goal to construct two satellites beside of rent the national SupremeSAT payload in Chinese satellites. Sri Lankan Telecommunications Regulatory Commission has signed an agreement with Surrey Satellite Technology Ltd to get relevant help and resources. Launch into Earth orbit would be done by a foreign provider.

- Syrian Space Research Center developing CubeSat-like small first national satellite since 2008.

- Tunisia is developing its first satellite, ERPSat01. Consisting of a CubeSat of 1 kg mass, it will be developed by the Sfax School of Engineering. ERPSat satellite is planned to be launched into orbit in 2013.

- Uzbekistan's State Space Research Agency (UzbekCosmos) announced in 2001 about intention of launch in 2002 first remote sensing satellite. Later in 2004 was stated that two satellites (remote sensing and telecommunication) will be built by Russia for $60–70 million each

## Attacks on Satellites

In recent times, satellites have been hacked by militant organizations to broadcast propaganda and to pilfer classified information from military communication networks.

For testing purposes, satellites in low earth orbit have been destroyed by ballistic missiles launched from earth. Russia, the United States and China have demonstrated the ability to eliminate satellites. In 2007 the Chinese military shot down an aging weather satellite, followed by the US Navy shooting down a defunct spy satellite in February 2008.

## Jamming

Due to the low received signal strength of satellite transmissions, they are prone to jamming by land-based transmitters. Such jamming is limited to the geographical area within the transmitter's range. GPS satellites are potential targets for jamming, but satellite phone and television signals have also been subjected to jamming.

Also, it is trivial to transmit a carrier radio signal to a geostationary satellite and thus interfere with the legitimate uses of the satellite's transponder. It is common for Earth stations to transmit at the wrong time or on the wrong frequency in commercial satellite space, and dual-illuminate the transponder, rendering the frequency unusable. Satellite operators now have sophisticated monitoring that enables them to pinpoint the source of any carrier and manage the transponder space effectively.

## Satellite Services

- Satellite crop monitoring
- Satellite Internet access
- Satellite navigation
- Satellite phone
- Satellite radio
- Satellite television

## Major Satellite Subsystems

## Satellite Bus

A satellite bus or spacecraft bus is the general model on which multiple-production satellite spacecraft are often based. The bus is the infrastructure of a spacecraft, usually providing locations for the payload (typically space experiments or instruments).

They are commonly used for geosynchronous satellites, particularly communications satellites, but are also used in spacecraft which occupy lower orbits, occasionally including low Earth orbit missions.

A bus-derived satellite would be used as opposed to a one-off, or specially produced satellite, such as Prospero X-3. Bus-derived satellites are usually customized to customer requirements, for example with specialized sensors or transponders, in order to achieve a specific mission.

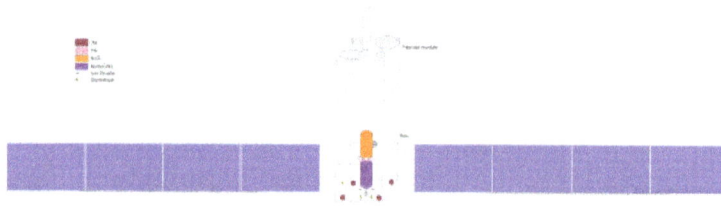

Communications satellite bus and payload module

## Fractionated Spacecraft

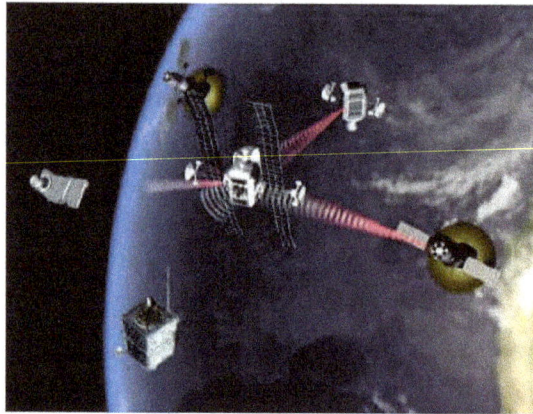

A fractionated spacecraft is a satellite architecture where the functional capabilities of a conventional monolithic spacecraft are distributed across multiple modules which interact through wireless links. Unlike other aggregations of spacecraft, such as constellations and clusters, the modules of a fractionated spacecraft are largely heterogeneous and perform distinct functions corresponding, for instance, to the various subsystem elements of a traditional satellite.

## History

The term "fractionated spacecraft" appears to have been coined by Owen Brown and Paul Eremenko in a series of 2006 papers, which argue that a fractionated architecture offers more flexibility and robustness than traditional satellite design during mission operations, and during the design and procurement.

The idea dates back to at least a 1984 article by P. Molette. Molette's, and later analyses by Rooney, concluded that the benefits of fractionated spacecraft were outweighed by their higher mass and cost. By 2006, Brown and his collaborators claim that the option value of flexibility, the insurance value of improved robustness, and mass production effects will exceed any penalties, and make an analogy with distributed clusters of personal computers (PCs) which are overtaking supercomputers. A 2006 study by the Massachusetts Institute of Technology appears to have corroborated this latter view.

# Development

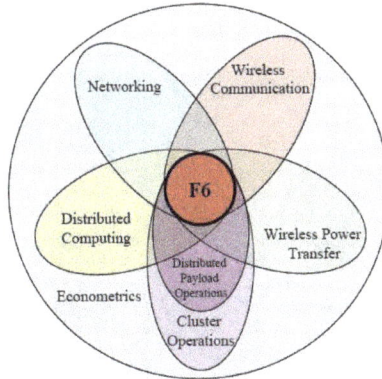

In 2007, DARPA, the Pentagon's advanced technology organization, issued an announcement soliciting proposals for a program entitled System F6, which aims to prove "the feasibility and benefits" of a fractionated satellite architecture through a space demonstration. The program appears to emphasize wireless networking as a critical technical enabler, along with econometric modeling to assess if and when the architecture is advantageous over conventional approaches

DARPA called for open source development of the networking and communications protocols and interfaces for the fractionated spacecraft modules. This unusual step was presumably in an effort to proliferate the concept and mirror in space the development of the terrestrial Internet.

In 2008, DARPA announced that contracts for the preliminary development phase of the System F6 program were issued to teams headed by Boeing, Lockheed Martin, Northrop Grumman, and Orbital Sciences. The second phase of the program was awarded to Orbital Sciences, along with IBM and JPL, in December 2009.

In May 16, 2013, DARPA confirmed that they cancelled the Formation-flying Satellite Demo, which means that they closed the project.

## Miscellaneous

Fractionating a communications satellite mission appears to be subject to U.S. Patent 6,633,745, "*Satellite cluster comprising a plurality of modular satellites*".

## Telemetry

A saltwater crocodile with a GPS-based satellite transmitter attached to its head for tracking

Telemetry is an automated communications process by which measurements and other data are collected at remote or inaccessible points and transmitted to receiving equipment for monitoring. The word is derived from Greek roots: *tele* = remote, and *metron* = measure. Systems that need external instructions and data to operate require the counterpart of telemetry, telecommand.

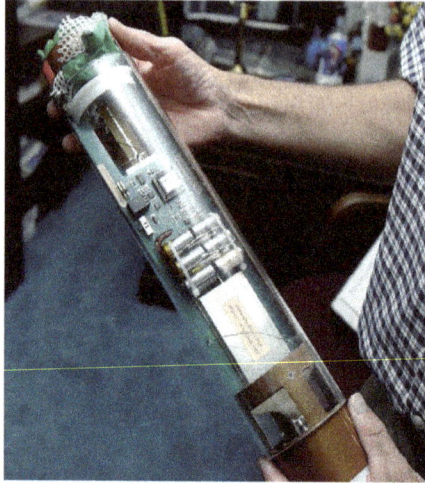

An expendable dropsonde used to capture weather data. The telemetry consists of sensors for pressure, temperature, and humidity and a wireless transmitter to return the captured data to an aircraft.

Although the term commonly refers to wireless data transfer mechanisms (e.g., using radio, ultrasonic, or infrared systems), it also encompasses data transferred over other media such as a telephone or computer network, optical link or other wired communications like phase line carriers. Many modern telemetry systems take advantage of the low cost and ubiquity of GSM networks by using SMS to receive and transmit telemetry data.

A telemeter is a device used to remotely measure any quantity. It consists of a sensor, a transmission path, and a display, recording, or control device. Telemeters are the physical devices used in telemetry. Electronic devices are widely used in telemetry and can be wireless or hard-wired, analog or digital. Other technologies are also possible, such as mechanical, hydraulic and optical.

Telemetry may be commutated to allow the transmission of multiple data streams in a fixed frame.

## History

Telemetering information over wire had its origins in the 19th century. One of the first data-transmission circuits was developed in 1845 between the Russian Tsar's Winter Palace and army headquarters. In 1874, French engineers built a system of weather and snow-depth sensors on Mont Blanc that transmitted real-time information to Paris. In 1901 the American inventor C. Michalke patented the selsyn, a circuit for sending synchronized rotation information over a distance. In 1906 a set of seismic stations were built with telemetering to the Pulkovo Observatory in Russia. In 1912, Commonwealth

Edison developed a system of telemetry to monitor electrical loads on its power grid. The Panama Canal (completed 1913–1914) used extensive telemetry systems to monitor locks and water levels.

Wireless telemetry made early appearances in the radiosonde, developed concurrently in 1930 by Robert Bureau in France and Pavel Molchanov in Russia. Mochanov's system modulated temperature and pressure measurements by converting them to wireless Morse code. The German V-2 rocket used a system of primitive multiplexed radio signals called "Messina" to report four rocket parameters, but it was so unreliable that Wernher von Braun once claimed it was more useful to watch the rocket through binoculars. In the US and the USSR, the Messina system was quickly replaced with better systems (in both cases, based on pulse-position modulation).

Early Soviet missile and space telemetry systems which were developed in the late 1940s used either pulse-position modulation (e.g., the Tral telemetry system developed by OKB-MEI) or pulse-duration modulation (e.g., the RTS-5 system developed by NII-885). In the United States, early work employed similar systems, but were later replaced by pulse-code modulation (PCM) (for example, in the Mars probe Mariner 4). Later Soviet interplanetary probes used redundant radio systems, transmitting telemetry by PCM on a decimeter band and PPM on a centimeter band.

## Applications

## Meteorology

Telemetry has been used by weather balloons for transmitting meteorological data since 1920.

## Oil and Gas Industry

Telemetry is used to transmit drilling mechanics and formation evaluation information uphole, in real time, as a well is drilled. These services are known as Measurement while drilling and Logging while drilling. Information acquired thousands of feet below ground, while drilling, is sent through the drilling hole to the surface sensors and the demodulation software. The pressure wave (sana) is translated into useful information after DSP and noise filters. This information is used for Formation evaluation, Drilling Optimization, and Geosteering.

## Motor Racing

Telemetry is a key factor in modern motor racing, allowing race engineers to interpret data collected during a test or race and use it to properly tune the car for optimum performance. Systems used in series such as Formula One have become advanced to the point where the potential lap time of the car can be calculated, and this time is what the driver is expected to meet. Examples of measurements on a race car include accelerations (G forces) in three axes, temperature readings, wheel speed, and suspension displacement. In Formula One, driver input is also recorded so the team can assess driver performance and (in case of an accident) the FIA can determine or rule out driver error as a possible cause.

Later developments include two-way telemetry which allows engineers to update calibrations on the car in real time (even while it is out on the track). In Formula One, two-way telemetry surfaced in the early 1990s and consisted of a message display on the dashboard which the team could update. Its development continued until May 2001, when it was first allowed on the cars. By 2002, teams were able to change engine mapping and deactivate engine sensors from the pit while the car was on the track. For the 2003 season, the FIA banned two-way telemetry from Formula One; however, the technology may be used in other types of racing or on road cars.

Telemetry has also been applied in yacht racing on Oracle Racing's USA 76.

One way telemetry system has also been applied in R/C racing car to get information by car's sensors like: engine RPM, voltage, temperatures, throttle.

## Transportation

In the transportation industry, telemetry provides meaningful information about the driver's performance by collecting data from the vehicle, leading to better fuel efficiency through driver feedback, which includes in-cab coaching. Other benefits include fewer traffic violations and lower insurance cost for trucking companies.

## Agriculture

Most activities related to healthy crops and good yields depend on timely availability of weather and soil data. Therefore, wireless weather stations play a major role in disease prevention and precision irrigation. These stations transmit parameters necessary for decision-making to a base station: air temperature and relative humidity, precipitation and leaf wetness (for disease prediction models), solar radiation and wind speed (to calculate evapotranspiration), water deficit stress (WDS) leaf sensors and soil moisture (crucial to irrigation decisions).

Because local micro-climates can vary significantly, such data needs to come from within the crop. Monitoring stations usually transmit data back by terrestrial radio, although occasionally satellite systems are used. Solar power is often employed to make the station independent of the power grid.

## Water Management

Telemetry is important in water management, including water quality and stream gauging functions. Major applications include AMR (automatic meter reading), groundwater monitoring, leak detection in distribution pipelines and equipment surveillance. Having data available in almost real time allows quick reactions to events in the field. Telemetry control allows to intervene with assets such as pumps and allows to remotely switch pumps on or off depending on the circumstances. Watershed telemetry is an excellent strategy of how to implement a water management system.

## Swimming Pools

Telemetry is used to transmit data in real time to server-based databases and applications with interfaces allowing monitoring and control. Server-side data storage and interpretation offers increased pool reliability. Additional data points, such as weather telemetry locally gathered or from Internet sources, can offer increased refinement of the control functions, reducing the requirement for consumables to manage water quality. Telemetry is also used to monitor health and usage of local equipment in the pump house. A properly designed telemetry and control system can offer additional benefits in professional management of maintenance contracts, with significant reduction in labor cost per client. PCFR SAS's PoolCop / PoolCopilot technology is an example of this approach to swimming pool management and control using telemetry transmitted from the pump house using the client's Internet service.

## Defense, Space and Resource Exploration

Telemetry is used in complex systems such as missiles, RPVs, spacecraft, oil rigs, and chemical plants since it allows the automatic monitoring, alerting, and record-keeping necessary for efficient and safe operation. Space agencies such as ISRO, NASA, the European Space Agency (ESA), and other agencies use telemetry and/or telecommand systems to collect data from spacecraft and satellites.

Telemetry is vital in the development of missiles, satellites and aircraft because the system might be destroyed during or after the test. Engineers need critical system parameters to analyze (and improve) the performance of the system. In the absence of telemetry, this data would often be unavailable.

## Space Science

Telemetry is used by manned or unmanned spacecraft for data transmission. Distances of more than 10 billion kilometres have been covered, e.g., by Voyager 1.

## Rocketry

In rocketry, telemetry equipment forms an integral part of the rocket range assets used to monitor the position and health of a launch vehicle to determine range safety flight termination criteria (Range purpose is for public safety). Problems include the extreme environment (temperature, acceleration and vibration), the energy supply, antenna alignment and (at long distances, e.g., in spaceflight) signal travel time.

## Flight Testing

Today nearly every type of aircraft, missiles, or spacecraft carries a wireless telemetry system as it is tested. Aeronautical mobile telemetry is used for the safety of the pilots and persons on the ground during flight tests. Telemetry from an on-board flight test instrumentation system is the primary source of real-time measurement and status information transmitted during the testing of manned and unmanned aircraft.

## Military Intelligence

Intercepted telemetry was an important source of intelligence for the United States and UK when Soviet missiles were tested; for this purpose, the United States operated a listening post in Iran. Eventually, the Russians discovered the United States intelligence-gathering network and encrypted their missile-test telemetry signals. Telemetry was also a source for the Soviets, who operated listening ships in Cardigan Bay to eavesdrop on UK missile tests performed in the area.

## Energy Monitoring

In factories, buildings and houses, energy consumption of systems such as HVAC are monitored at multiple locations; related parameters (e.g., temperature) are sent via wireless telemetry to a central location. The information is collected and processed, enabling the most efficient use of energy. Such systems also facilitate predictive maintenance.

## Resource Distribution

Many resources need to be distributed over wide areas. Telemetry is useful in these cases, since it allows the system to channel resources where they are needed; examples of this are tank farms in gasoline refineries and chemical plants.

## Medicine/Healthcare

Telemetry is used for patients (biotelemetry) who are at risk of abnormal heart activity, generally in a coronary care unit. Telemetry specialists are sometimes used to monitor many patients with a hospital. Such patients are outfitted with measuring, recording and transmitting devices. A data log can be useful in diagnosis of the patient's condition by doctors. An alerting function can alert nurses if the patient is suffering from an acute (or dangerous) condition.

Systems are available in medical-surgical nursing for monitoring to rule out a heart condition, or to monitor a response to antiarrhythmic medications such as amiodarone.

A new and emerging application for telemetry is in the field of neurophysiology, or neurotelemetry. Neurophysiology is the study of the central and peripheral nervous systems through the recording of bioelectrical activity, whether spontaneous or stimulated. In neurotelemetry (NT) the electroencephalogram (EEG) of a patient is monitored remotely by a registered EEG technologist using advanced communication software. The goal of neurotelemetry is to recognize a decline in a patient's condition before physical signs and symptoms are present.

Neurotelemetry is synonymous with real-time continuous video EEG monitoring and has application in the epilepsy monitoring unit, neuro ICU, pediatric ICU and newborn ICU. Due to the labor-intensive nature of continuous EEG monitoring NT is typically done in the larger academic teaching hospitals using in-house programs that include R.EEG Technologists, IT support staff, neurologist and neurophysiologist and monitoring support personnel.

Modern microprocessor speeds, software algorithms and video data compression allow hospitals to centrally record and monitor continuous digital EEGs of multiple critically ill patients simultaneously.

Neurotelemetry and continuous EEG monitoring provides dynamic information about brain function that permits early detection of changes in neurologic status, which is especially useful when the clinical examination is limited.

## Fishery and Wildlife Research and Management

Telemetry is used to study wildlife, and has been useful for monitoring threatened species at the individual level. Animals under study can be outfitted with instrumentation tags, which include sensors that measure temperature, diving depth and duration (for marine animals), speed and location (using GPS or Argos packages). Telemetry tags can give researchers information about animal behavior, functions, and their environment. This information is then either stored (with archival tags) or the tags can send (or transmit) their information to a satellite or handheld receiving device. Capturing and marking wild animals can put them at some risk, so it is important to minimize these impacts.

## Retail

At a 2005 workshop in Las Vegas, a seminar noted the introduction of telemetry equipment which would allow vending machines to communicate sales and inventory data to a route truck or to a headquarters. This data could be used for a variety of purposes, such as eliminating the need for drivers to make a first trip to see which items needed to be restocked before delivering the inventory.

Retailers also use RFID tags to track inventory and prevent shoplifting. Most of these tags passively respond to RFID readers (e.g., at the cashier), but active RFID tags are available which periodically transmit location information to a base station.

## Law Enforcement

Telemetry hardware is useful for tracking persons and property in law enforcement. An ankle collar worn by convicts on probation can warn authorities if a person violates the terms of his or her parole, such as by straying from authorized boundaries or visiting an unauthorized location. Telemetry has also enabled bait cars, where law enforcement can rig a car with cameras and tracking equipment and leave it somewhere they expect it to be stolen. When stolen the telemetry equipment reports the location of the vehicle, enabling law enforcement to deactivate the engine and lock the doors when it is stopped by responding officers.

## Energy Providers

In some countries, telemetry is used to measure the amount of electrical energy consumed. The electricity meter communicates with a concentrator, and the latter sends the information through GPRS or GSM to the energy provider's server. Telemetry is also used for the remote monitoring of substations and their equipment. For data transmission, phase line carrier systems operating on frequencies between 30 and 400 kHz are sometimes used.

## Falconry

In falconry, "telemetry" means a small radio transmitter carried by a bird of prey that will allow the bird's owner to track it when it is out of sight.

## Testing

Telemetry is used in testing hostile environments which are dangerous for humans to be present . Examples include munitions storage facilities, radioactive sites, volcanoes, deep sea, and outer space

## Communications

Telemetry is used in many battery operated wireless systems to inform monitoring personnel when the battery power is reaching a low point and the end item needs fresh batteries.

## Mining

In the mining industry, telemetry serves two main purposes: the measurement of key parameters from mining equipment and the monitoring of safety practices. The information provided by the collection and analysis of key parameters allows for root-cause identification of inefficient operations, unsafe practices and incorrect equipment usage for maximizing productivity and safety. Further applications of the technology allow for sharing knowledge and best practices across the organization.

## International Standards

As in other telecommunications fields, international standards exist for telemetry equipment and software. CCSDS and IRIG are such standards.

## GNSS Software-defined Receiver

A software GNSS receiver is a GNSS receiver that has been designed and implemented following the philosophy of Software-defined radio.

A GNSS receiver, in general, is an electronic device that receives and digitally processes the signals from a GNSS satellite constellation in order to provide position, velocity and time (of the receiver).

GNSS receivers have been traditionally implemented in hardware: a *hardware GNSS receiver* is conceived as a dedicated chip that have been designed and built (from the very beginning) with the only purpose of being a GNSS receiver.

In a software GNSS receiver, all digital processing is performed by a general purpose microprocessor. In this approach, a small amount of inexpensive hardware is still needed, known as the *frontend*, that digitizes the signal from the satellites. The microprocessor can then work on this *raw* digital stream to implement the GNSS functionality.

## Hardware vs. Software GNSS Receivers

When comparing *hardware* vs *software* GNSS receivers, a number of pros and cons can be found for each approach:

- Hardware GNSS receivers are in general more efficient from the point of view of both computational load and power consumption since they have been designed in a highly specialized way with the only purpose of implementing the GNSS processing.

- Software GNSS receivers allow a huge flexibility: many features of the receiver can be modified just through software. This provides the receiver with adaptive capabilities, depending on the user's needs and working conditions. In addition, the receiver can be easily upgraded via software.

- Under some assumptions, Software GNSS receivers can be more profitable for some applications, as long as sufficient computational power is available (and can be shared among multiple applications). For example, the microprocessor of a smartphone can be used to provide GNSS navigation with the only need of including a frontend (instead of a full, more expensive, hardware receiver).

Currently, most of the GNSS receiver market is still *hardware*. However, there already exist operational solutions based on the software approach able to run on low-cost microprocessors. Software GNSS receivers are expected to increase their market share or even take over in the near future, following the development of the computational capabilities of the microprocessors (Moore's law).

## References

- Bleiler, Everett Franklin; Bleiler, Richard (1991). Science-fiction, the Early Years. Kent State University Press. p. 325. ISBN 978-0-87338-416-2.

- Grant, A.; Meadows, J. (2004). Communication Technology Update (ninth ed.). Focal Press. p. 284. ISBN 0-240-80640-9.

- Burleson, Daphne (2005). Space Programs Outside the United States. McFarland & Company. p. 43. ISBN 978-0-7864-1852-7.

- Gruntman, Mike (2004). Blazing the Trail. American Institute of Aeronautics and Astronautics. p. 426. ISBN 978-1-56347-705-8.

- Harvey, Brian (2003). Europe's Space Programme. Springer Science+Business Media. p. 114. ISBN 978-1-85233-722-3.

- Telemetry in the Mining Industry. IETE Journal of Research. Volume 29, Issue 8, 1983. Retrieved August 20th 2015.

- Operational Safety and Efficiency of Mobile Equipment through Operator Behavior Monitoring. Canadian Institute of Mining. 2015.

- "Antrix Corporation Ltd - Satellites > Spacecraft Systems & Sub Systems". Antrix.gov.in. 2009-09-24. Retrieved 2014-04-23.

- "Telemetry: Summary of concept and rationale". NASA report. SAO/NASA ADS Physics Abstract Service. Retrieved 19 December 2014.

- La télémétrie véhiculaire, une technologie incontournable. Bernard Gauthier, Transport Magazine. June 2014.

- Rising, David (11 November 2013). "Satellite hits Atlantic — but what about next one?". Seattle Times. Archived from the original on November 12, 2013.

- "Global Experts Agree Action Needed on Space Debris". European Space Agency. 25 April 2013.

- "Nanosatellite Launch Service". University of Toronto Institute for Aerospace Studies. Archived from the original on March 10, 2013. Retrieved 2013-03-02.

- Italian built (by La Sapienza) first Iraqi small experimental Earth observation cubesat-satellite Tigrisat Iraq to launch its first satellite before the end of 2013 launched in 2014

- Bray, Allison (1 December 2012). "Students hope to launch first ever Irish satellite". The Independent. Ireland.

- "Burma to launch first state-owned satellite, expand communications". News. Mizzima. 14 June 2011. Archived from the original on 2011-06-17.

# Types of Satellites

Satellites can be categorized by their various applications in space telescopes, communications satellite, Earth observation satellites, weather satellites, reconnaissance satellites and biosatellites. The chapter provides an in-depth analysis of each type with suitable examples to enable better understanding.

## Space Telescope

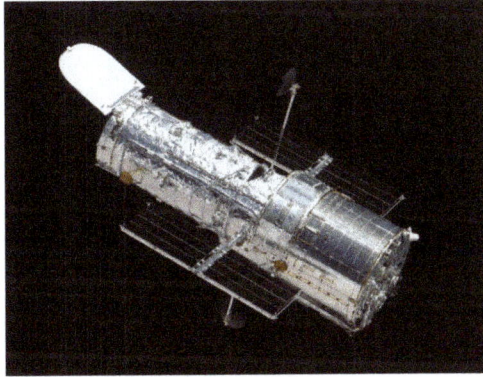

The Hubble Space Telescope, one of the Great Observatories

A space telescope or space observatory is an instrument located in outer space to observe distant planets, galaxies and other astronomical objects. Space telescopes avoid many of the problems of ground-based observatories, such as light pollution and distortion of electromagnetic radiation (scintillation). In addition, ultraviolet frequencies, X-rays and gamma rays are blocked by the Earth's atmosphere, so they can only be observed from space.

Theorized by Lyman Spitzer in 1946, the first operational space telescopes were the American Orbiting Astronomical Observatory OAO-2 launched in 1968 and the Soviet Orion 1 ultraviolet telescope aboard space station Salyut 1 in 1971.

Space telescopes are distinct from other imaging satellites pointed toward Earth for purposes of espionage, weather analysis and other types of information gathering.

### History

In 1946, American theoretical astrophysicist Lyman Spitzer was the first to conceive the idea of a telescope in outer space, 11 years before the Soviet Union launched the first satellite, *Sputnik 1*, in October 1957.

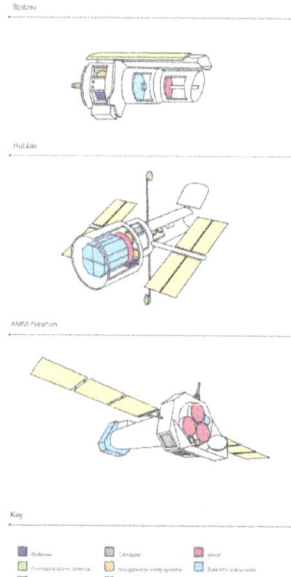

Spitzer, Hubble and XMM with their most important parts depicted

Spitzer's proposal called for a large telescope that would not be hindered by Earth's atmosphere. After lobbying in the 1960s and 70s for such a system to be built, Spitzer's vision ultimately materialized into the Hubble Space Telescope, which was launched on April 24, 1990 by the Space Shuttle *Discovery* (STS-31).

## Advantages

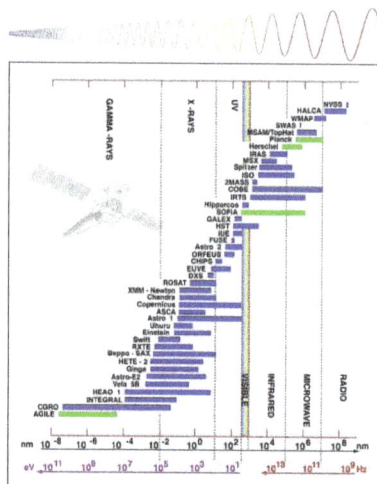

Space observatories and their wavelength working ranges

Performing astronomy from ground-based observatories on Earth is limited by the filtering and distortion of electromagnetic radiation (scintillation or twinkling) due to the atmosphere. Some terrestrial telescopes can reduce atmospheric effects with adaptive optics. A telescope orbiting Earth outside the atmosphere is subject neither to twinkling nor to light pollution from artificial light sources on Earth. As a result, the angular resolution of space telescopes is often much smaller than a ground-based telescope with a similar aperture.

Space-based astronomy is even more important for frequency ranges which are outside the optical window and the radio window, the only two wavelength ranges of the electromagnetic spectrum that are not severely attenuated by the atmosphere. For example, X-ray astronomy is nearly impossible when done from Earth, and has reached its current importance in astronomy only due to orbiting X-ray telescopes such as the Chandra observatory and the XMM-Newton observatory. Infrared and ultraviolet are also largely blocked.

However, all these advantages do come with a price. Space telescopes are much more expensive to build than ground-based telescopes. Due to their location, space telescopes are also extremely difficult to maintain. The Hubble Space Telescope was serviced by the Space Shuttle while many other space telescopes cannot be serviced at all.

Space observatories can generally be divided into two classes: missions which map the entire sky (surveys), and observatories which focus on selected astronomical objects or parts of the sky.

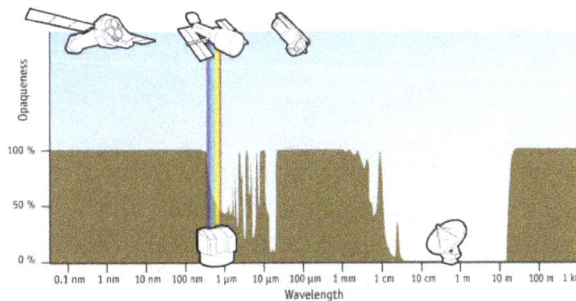

Space observatories and their wavelength working range

Many space observatories have already completed their missions, while others continue operating, and still others are planned for the future. Satellites have been launched and operated by NASA, ISRO, ESA, Japanese Space Agency and the Soviet space program later succeeded by Roskosmos of Russia.

# Communications Satellite

A communications satellite is an artificial satellite that relays and amplifies radio telecommunications signals via a transponder; it creates a communication channel between a source transmitter and a receiver at different locations on Earth. Communications satellites are used for television, telephone, radio, internet, and military applications. There are over 2,000 communications satellites in Earth's orbit, used by both private and government organizations.

An Advanced Extremely High Frequency communications satellite
relays secure communications for the United States and other allied countries.

Wireless communication uses electromagnetic waves to carry signals. These waves require line-of-sight, and are thus obstructed by the curvature of the Earth. The purpose of communications satellites is to relay the signal around the curve of the Earth allowing communication between widely separated points. Communications satellites use a wide range of radio and microwave frequencies. To avoid signal interference, international organizations have regulations for which frequency ranges or "bands" certain organizations are allowed to use. This allocation of bands minimizes the risk of signal interference.

## History

The concept of the geostationary communications satellite was first proposed by Arthur C. Clarke, building on work by Konstantin Tsiolkovsky and on the 1929 work by Herman Potočnik (writing as Herman Noordung) *Das Problem der Befahrung des Weltraums — der Raketen-motor*. In October 1945 Clarke published an article titled "Extraterrestrial Relays" in the British magazine *Wireless World*. The article described the fundamentals behind the deployment of artificial satellites in geostationary orbits for the purpose of relaying radio signals. Thus, Arthur C. Clarke is often quoted as being the inventor of the communications satellite and the term 'Clarke Belt' employed as a description of the orbit.

Decades later a project named Communication Moon Relay was a telecommunication project carried out by the United States Navy. Its objective was to develop a secure and reliable method of wireless communication by using the Moon as a passive reflector and natural communications satellite.

The first artificial Earth satellite was Sputnik 1. Put into orbit by the Soviet Union on October 4, 1957, it was equipped with an on-board radio-transmitter that worked on two frequencies: 20.005 and 40.002 MHz. Sputnik 1 was launched as a step in the exploration of space and rocket development. While incredibly important it was not placed in orbit for the purpose of sending data from one point on earth to another. And it was the first artificial satellite in the steps leading to today's satellite communications.

The first artificial satellite used solely to further advances in global communications was a balloon named Echo 1. Echo 1 was the world's first artificial communications satellite capable of relaying signals to other points on Earth. It soared 1,600 kilometres (1,000 mi) above the planet after its Aug. 12, 1960 launch, yet relied on humanity's oldest flight technology — ballooning. Launched by NASA, Echo 1 was a 30-metre (100 ft) aluminised PET film balloon that served as a passive reflector for radio communications. The world's first inflatable satellite — or "satelloon", as they were informally known — helped lay the foundation of today's satellite communications. The idea behind a communications satellite is simple: Send data up into space and beam it back down to another spot on the globe. Echo 1 accomplished this by essentially serving as an enormous mirror, 10 stories tall, that could be used to reflect communications signals.

The first American satellite to relay communications was Project SCORE in 1958, which used a tape recorder to store and forward voice messages. It was used to send a Christmas greeting to the world from U.S. President Dwight D. Eisenhower.; Courier 1B, built by Philco, launched in 1960, was the world's first active repeater satellite.

There are two major classes of communications satellites, *passive* and *active*. Passive satellites only reflect the signal coming from the source, toward the direction of the receiver. With passive satellites, the reflected signal is not amplified at the satellite, and only a very small amount of the transmitted energy actually reaches the receiver. Since the satellite is so far above Earth, the radio signal is attenuated due to free-space path loss, so the signal received on Earth is very weak. Active satellites, on the other hand, amplify the received signal before re-transmitting it to the receiver on the ground. Passive satellites were the first communications satellites, but are little used now. Telstar was the second active, direct relay communications satellite. Belonging to AT&T as part of a multi-national agreement between AT&T, Bell Telephone Laboratories, NASA, the British General Post Office, and the French National PTT (Post Office) to develop satellite communications, it was launched by NASA from Cape Canaveral on July 10, 1962, the first privately sponsored space launch. Relay 1 was launched on December 13, 1962, and became the first satellite to broadcast across the Pacific on November 22, 1963.

An immediate antecedent of the geostationary satellites was Hughes' Syncom 2, launched on July 26, 1963. Syncom 2 was the first communications satellite in a geosynchronous orbit. It revolved around the earth once per day at constant speed, but because it still had north-south motion, special equipment was needed to track it. Its successor, Syncom 3 was the first geostationary communications satellite. Syncom 3 obtained a geosynchronous orbit, without a north-south motion, making it appear from the ground as a stationary object in the sky.

Beginning with the Mars Exploration Rovers, probes on the surface of Mars have used orbiting spacecraft as communications satellites for relaying their data to Earth. The orbiters were designed for this relay purpose to allow the landers to conserve power. The Orbiters with their solar power arrays, large antennas and more powerful transmitters enable them to transmit data to Earth with a much stronger, and as a result, clearer signal than a lander could manage on its own from the surface.

## Satellite Orbits

Communications satellites usually have one of three primary types of orbit, while other orbital classifications are used to further specify orbital details.

- Geostationary satellites have a *geostationary orbit* (GEO), which is 35,786 kilometres (22,236 mi) from Earth's surface. This orbit has the special characteristic that the apparent position of the satellite in the sky when viewed by a ground observer does not change, the satellite appears to "stand still" in the sky. This is because the satellite's orbital period is the same as the rotation rate of the Earth. The advantage of this orbit is that ground antennas do not have to track the satellite across the sky, they can be fixed to point at the location in the sky the satellite appears.

- *Medium Earth orbit* (MEO) satellites are closer to Earth. Orbital altitudes range from 2,000 to 35,786 kilometres (1,243 to 22,236 mi) above Earth.

- The region below medium orbits is referred to as *low Earth orbit* (LEO), and is about 160 to 2,000 kilometres (99 to 1,243 mi) above Earth.

As satellites in MEO and LEO orbit the Earth faster, they do not remain visible in the sky to a fixed point on Earth continually like a geostationary satellite, but appear to a ground observer to cross the sky and "set" when they go behind the Earth. Therefore, to provide continuous communications capability with these lower orbits requires a larger number of satellites, so one will always be in the sky for transmission of communication signals. However, due to their relatively small distance to the Earth their signals are stronger.

## Low Earth Orbiting (LEO) Satellites

A low Earth orbit (LEO) typically is a circular orbit about 160 to 2,000 kilometres (99 to 1,243 mi) above the earth's surface and, correspondingly, a period (time to revolve around the earth) of about 90 minutes.

Because of their low altitude, these satellites are only visible from within a radius of roughly 1,000 kilometres (620 mi) from the sub-satellite point. In addition, satellites in low earth orbit change their position relative to the ground position quickly. So even for local applications, a large number of satellites are needed if the mission requires uninterrupted connectivity.

Low-Earth-orbiting satellites are less expensive to launch into orbit than geostationary satellites and, due to proximity to the ground, do not require as high signal strength (*Recall that signal strength falls off as the square of the distance from the source, so the effect is dramatic*). Thus there is a trade off between the number of satellites and their cost.

In addition, there are important differences in the onboard and ground equipment needed to support the two types of missions.

## Satellite Constellation

A group of satellites working in concert is known as a satellite constellation. Two such constellations, intended to provide satellite phone services, primarily to remote areas, are the Iridium and Globalstar systems. The Iridium system has 66 satellites.

It is also possible to offer discontinuous coverage using a low-Earth-orbit satellite capable of storing data received while passing over one part of Earth and transmitting it later while passing over another part. This will be the case with the CASCADE system of Canada's CASSIOPE communications satellite. Another system using this store and forward method is Orbcomm.

## Medium Earth Orbit (MEO)

A MEO is a satellite in orbit somewhere between 2,000 and 35,786 kilometres (1,243 and 22,236 mi) above the earth's surface. MEO satellites are similar to LEO satellites in functionality. MEO satellites are visible for much longer periods of time than LEO satellites, usually between 2 and 8 hours. MEO satellites have a larger coverage area than LEO satellites. A MEO satellite's longer duration of visibility and wider footprint means fewer satellites are needed in a MEO network than a LEO network. One disadvantage is that a MEO satellite's distance gives it a longer time delay and weaker signal than a LEO satellite, although these limitations are not as severe as those of a GEO satellite.

Like LEOs, these satellites don't maintain a stationary distance from the earth. This is in contrast to the geostationary orbit, where satellites are always approximately 35,786 kilometres (22,236 mi) from the earth.

Typically the orbit of a medium earth orbit satellite is about 16,000 kilometres (10,000 mi) above earth. In various patterns, these satellites make the trip around earth in anywhere from 2–12 hours, which provides better coverage to wider areas than that provided by LEOs.

## Example

In 1962, the first communications satellite, Telstar, was launched. It was a medium earth orbit satellite designed to help facilitate high-speed telephone signals. Although it was the first practical way to transmit signals over the horizon, its major drawback was soon realized. Because its orbital period of about 2.5 hours did not match the Earth's rotational period of 24 hours, continuous coverage was impossible. It was apparent that multiple MEOs needed to be used in order to provide continuous coverage.

## Geostationary Orbits (GEO)

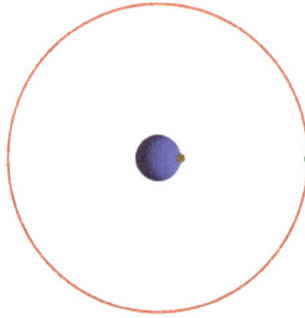

Geostationary orbit

To an observer on the earth, a satellite in a geostationary orbit appears motionless, in a fixed position in the sky. This is because it revolves around the earth at the earth's own angular velocity (360 degrees every 24 hours, in an equatorial orbit).

A geostationary orbit is useful for communications because ground antennas can be aimed at the satellite without their having to track the satellite's motion. This is relatively inexpensive.

In applications that require a large number of ground antennas, such as DirecTV distribution, the savings in ground equipment can more than outweigh the cost and complexity of placing a satellite into orbit.

## Examples

- The first geostationary satellite was Syncom 3, launched on August 19, 1964, and used for communication across the Pacific starting with television coverage of the 1964 Summer Olympics. Shortly after Syncom 3, Intelsat I, aka *Early Bird*, was launched on April 6, 1965 and placed in orbit at 28° west longitude. It was the first geostationary satellite for telecommunications over the Atlantic Ocean.

- On November 9, 1972, Canada's first geostationary satellite serving the continent, Anik A1, was launched by Telesat Canada, with the United States following suit with the launch of Westar 1 by Western Union on April 13, 1974.

- On May 30, 1974, the first geostationary communications satellite in the world to be three-axis stabilized was launched: the experimental satellite ATS-6 built for NASA.

- After the launches of the Telstar through Westar 1 satellites, RCA Americom (later GE Americom, now SES) launched Satcom 1 in 1975. It was Satcom 1 that was instrumental in helping early cable TV channels such as WTBS (now TBS Superstation), HBO, CBN (now ABC Family) and The Weather Channel become successful, because these channels distributed their programming to all of the local cable TV headends using the satellite. Additionally, it was the first satellite used by broadcast television networks in the United States, like ABC, NBC, and CBS, to distribute programming to their local affiliate stations. Satcom 1 was widely used because it had twice the communications capacity of the competing Westar 1 in America (24 transponders as opposed to the 12 of Westar 1), resulting in lower transponder-usage costs. Satellites in later decades tended to have even higher transponder numbers.

By 2000, Hughes Space and Communications (now Boeing Satellite Development Center) had built nearly 40 percent of the more than one hundred satellites in service worldwide. Other major satellite manufacturers include Space Systems/Loral, Orbital Sciences Corporation with the STAR Bus series, Indian Space Research Organisation, Lockheed Martin (owns the former RCA Astro Electronics/GE Astro Space business), Northrop Grumman, Alcatel Space, now Thales Alenia Space, with the Spacebus series, and Astrium.

## Molniya Satellites

Geostationary satellites must operate above the equator and therefore appear lower on the horizon as the receiver gets the farther from the equator. This will cause problems for extreme northerly latitudes, affecting connectivity and causing multipath interference (caused by signals reflecting off the ground and into the ground antenna).

Thus, for areas close to the North (and South) Pole, a geostationary satellite may appear below the horizon. Therefore, Molniya orbit satellites have been launched, mainly in Russia, to alleviate this problem.

Molniya orbits can be an appealing alternative in such cases. The Molniya orbit is highly inclined, guaranteeing good elevation over selected positions during the northern portion of the orbit. (Elevation is the extent of the satellite's position above the horizon. Thus, a satellite at the horizon has zero elevation and a satellite directly overhead has elevation of 90 degrees.)

The Molniya orbit is designed so that the satellite spends the great majority of its time over the far northern latitudes, during which its ground footprint moves only slightly. Its period is one half day, so that the satellite is available for operation over the targeted region for six to nine hours every second revolution. In this way a constellation of three Molniya satellites (plus in-orbit spares) can provide uninterrupted coverage.

The first satellite of the Molniya series was launched on April 23, 1965 and was used for experimental transmission of TV signals from a Moscow uplink station to downlink stations located in Siberia and the Russian Far East, in Norilsk, Khabarovsk, Magadan and Vladivostok. In November 1967 Soviet engineers created a unique system of national TV network of satellite television, called Orbita, that was based on Molniya satellites.

## Polar Orbit

In the United States, the National Polar-orbiting Operational Environmental Satellite System (NPOESS) was established in 1994 to consolidate the polar satellite operations of NASA (National Aeronautics and Space Administration) NOAA (National Oceanic and Atmospheric Administration). NPOESS manages a number of satellites for various purposes; for example, METSAT for meteorological satellite, EUMETSAT for the European branch of the program, and METOP for meteorological operations.

These orbits are sun synchronous, meaning that they cross the equator at the same local time each day. For example, the satellites in the NPOESS (civilian) orbit will cross the equator, going from south to north, at times 1:30 P.M., 5:30 P.M., and 9:30 P.M.

## Structure

Communications Satellites are usually composed of the following subsystems:

- Communication Payload, normally composed of transponders, antennas, and switching systems

- Engines used to bring the satellite to its desired orbit

- Station Keeping Tracking and stabilization subsystem used to keep the satellite in the right orbit, with its antennas pointed in the right direction, and its power system pointed towards the sun

- Power subsystem, used to power the Satellite systems, normally composed of solar cells, and batteries that maintain power during solar eclipse

- Command and Control subsystem, which maintains communications with ground control stations. The ground control earth stations monitor the satellite performance and control its functionality during various phases of its life-cycle.

The bandwidth available from a satellite depends upon the number of transponders provided by the satellite. Each service (TV, Voice, Internet, radio) requires a different amount of bandwidth for transmission. This is typically known as link budgeting and a network simulator can be used to arrive at the exact value.

## Frequency Allocation for Satellite Systems

Allocating frequencies to satellite services is a complicated process which requires international coordination and planning. This is carried out under the auspices of the International Telecommunication Union (ITU). To facilitate frequency planning, the world is divided into three regions: Region 1: Europe, Africa, what was formerly the Soviet Union, and Mongolia Region 2: North and South America and Greenland Region 3: Asia (excluding region 1 areas), Australia, and the southwest Pacific

Within these regions, frequency bands are allocated to various satellite services, although a given service may be allocated different frequency bands in different regions. Some of the services provided by satellites are:

- Fixed satellite service (FSS)

- Broadcasting satellite service (BSS)

- Mobile satellite service

- Radionavigation-satellite service

- Meteorological-satellite service

- Amateur-satellite service

## Applications

## Telephone

An Iridium satellite

The first and historically most important application for communication satellites was in intercontinental long distance telephony. The fixed Public Switched Telephone Network relays telephone calls from land line telephones to an earth station, where they are then transmitted to a geostationary satellite. The downlink follows an analogous path. Improvements in submarine communications cables through the use of fiber-optics caused some decline in the use of satellites for fixed telephony in the late 20th century.

Satellite communications are still used in many applications today. Remote islands such as Ascension Island, Saint Helena, Diego Garcia, and Easter Island, where no submarine cables are in service, need satellite telephones. There are also regions of some continents and countries where landline telecommunications are rare to nonexistent, for example large regions of South America, Africa, Canada, China, Russia, and Australia. Satellite communications also provide connection to the edges of Antarctica and Greenland. Other land use for satellite phones are rigs at sea, a back up for hospitals, military, and recreation. Ships at sea, as well as planes, often use satellite phones.

Satellite phone systems can be accomplished by a number of means. On a large scale, often there will be a local telephone system in an isolated area with a link to the telephone system in a main land area. There are also services that will patch a radio signal to a telephone system. In this example, almost any type of satellite can be used. Satellite phones connect directly to a constellation of either geostationary or low-earth-orbit satellites. Calls are then forwarded to a satellite teleport connected to the Public Switched Telephone Network .

## Television

As television became the main market, its demand for simultaneous delivery of relatively few signals of large bandwidth to many receivers being a more precise match for the capabilities of geosynchronous comsats. Two satellite types are used for North American television and radio: Direct broadcast satellite (DBS), and Fixed Service Satellite (FSS).

The definitions of FSS and DBS satellites outside of North America, especially in Europe, are a bit more ambiguous. Most satellites used for direct-to-home television in Europe have the same high power output as DBS-class satellites in North America, but use the same linear polarization as FSS-class satellites. Examples of these are the Astra, Eutelsat, and Hotbird spacecraft in orbit over the European continent. Because of this, the terms FSS and DBS are more so used throughout the North American continent, and are uncommon in Europe.

Fixed Service Satellites use the C band, and the lower portions of the $K_u$ band. They are normally used for broadcast feeds to and from television networks and local affiliate stations (such as program feeds for network and syndicated programming, live shots, and backhauls), as well as being used for distance learning by schools and universities, business television (BTV), Videoconferencing, and general commercial telecommunications. FSS satellites are also used to distribute national cable channels to cable television headends.

Free-to-air satellite TV channels are also usually distributed on FSS satellites in the $K_u$ band. The Intelsat Americas 5, Galaxy 10R and AMC 3 satellites over North America provide a quite large amount of FTA channels on their $K_u$ band transponders.

The American Dish Network DBS service has also recently utilized FSS technology as well for their programming packages requiring their SuperDish antenna, due to Dish Network needing more capacity to carry local television stations per the FCC's "must-carry" regulations, and for more bandwidth to carry HDTV channels.

A direct broadcast satellite is a communications satellite that transmits to small DBS satellite dishes (usually 18 to 24 inches or 45 to 60 cm in diameter). Direct broadcast satellites generally operate in the upper portion of the microwave $K_u$ band. DBS technology is used for DTH-oriented (Direct-To-Home) satellite TV services, such as DirecTV and DISH Network in the United States, Bell TV and Shaw Direct in Canada, Freesat and Sky in the UK, Ireland, and New Zealand and DSTV in South Africa.

Operating at lower frequency and lower power than DBS, FSS satellites require a much larger dish for reception (3 to 8 feet (1 to 2.5 m) in diameter for $K_u$ band, and 12 feet (3.6 m) or larger for C band). They use linear polarization for each of the transponders' RF input and output (as opposed to circular polarization used by DBS satellites), but this is a minor technical difference that users do not notice. FSS satellite technology was also originally used for DTH satellite TV from the late 1970s to the early 1990s in the United States in the form of TVRO (TeleVision Receive Only) receivers and dishes. It was also used in its $K_u$ band form for the now-defunct Primestar satellite TV service.

Some satellites have been launched that have transponders in the $K_a$ band, such as DirecTV's SPACEWAY-1 satellite, and Anik F2. NASA and ISRO have also launched experimental satellites carrying $K_a$ band beacons recently.

Some manufacturers have also introduced special antennas for mobile reception of DBS television. Using Global Positioning System (GPS) technology as a reference, these antennas automatically re-aim to the satellite no matter where or how the vehicle (on which the antenna is mounted) is situated. These mobile satellite antennas are popular with some recreational vehicle owners. Such mobile DBS antennas are also used by JetBlue Airways for DirecTV (supplied by LiveTV, a subsidiary of JetBlue), which passengers can view on-board on LCD screens mounted in the seats.

## Radio Broadcasting

Satellite radio offers audio broadcast services in some countries, notably the United States. Mobile services allow listeners to roam a continent, listening to the same audio programming anywhere.

A satellite radio or subscription radio (SR) is a digital radio signal that is broadcast by a communications satellite, which covers a much wider geographical range than terrestrial radio signals.

Satellite radio offers a meaningful alternative to ground-based radio services in some countries, notably the United States. Mobile services, such as SiriusXM, and Worldspace, allow listeners to roam across an entire continent, listening to the same audio programming anywhere they go. Other services, such as Music Choice or Muzak's satellite-delivered content, require a fixed-location receiver and a dish antenna. In all cases, the antenna must have a clear view to the satellites. In areas where tall buildings, bridges, or even parking garages obscure the signal, repeaters can be placed to make the signal available to listeners.

Initially available for broadcast to stationary TV receivers, by 2004 popular mobile direct broadcast applications made their appearance with the arrival of two satellite radio systems in the United States: Sirius and XM Satellite Radio Holdings. Later they merged to become the conglomerate SiriusXM.

Radio services are usually provided by commercial ventures and are subscription-based. The various services are proprietary signals, requiring specialized hardware for decoding and playback. Providers usually carry a variety of news, weather, sports, and music channels, with the music channels generally being commercial-free.

In areas with a relatively high population density, it is easier and less expensive to reach the bulk of the population with terrestrial broadcasts. Thus in the UK and some other countries, the contemporary evolution of radio services is focused on Digital Audio Broadcasting (DAB) services or HD Radio, rather than satellite radio.

## Amateur Radio

Amateur radio operators have access to amateur satellites, which have been designed specifically to carry amateur radio traffic. Most such satellites operate as spaceborne repeaters, and are generally accessed by amateurs equipped with UHF or VHF radio equipment and highly directional antennas such as Yagis or dish antennas. Due to launch costs, most current amateur satellites are launched into fairly low Earth orbits, and are designed to deal with only a limited number of brief contacts at any given time. Some satellites also provide data-forwarding services using the X.25 or similar protocols.

## Internet Access

After the 1990s, satellite communication technology has been used as a means to connect to the Internet via broadband data connections. This can be very useful for users who are located in remote areas, and cannot access a broadband connection, or require high availability of services.

## Military

Communications satellites are used for military communications applications, such as Global Command and Control Systems. Examples of military systems that use communication satellites are the MILSTAR, the DSCS, and the FLTSATCOM of the United States, NATO satellites, United Kingdom satellites (for instance Skynet), and satellites of the former Soviet Union. India has launched its first Military Communication satellite GSAT-7, its transponders operate in UHF, F, C and $K_u$ band bands. Typically military satellites operate in the UHF, SHF (also known as X-band) or EHF (also known as $K_a$ band) frequency bands.

# Earth Observation Satellite

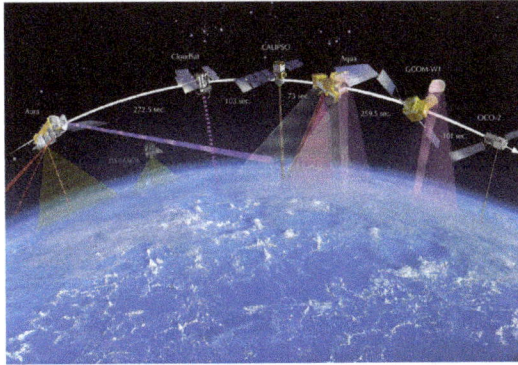

Six Earth observation satellites comprising the A-train satellite constellation as of 2014.

Earth observation satellites are satellites specifically designed for Earth observation from orbit, similar to spy satellites but intended for non-military uses such as environmental monitoring, meteorology, map making etc.

Most Earth observation satellites carry instruments that should be operated at a relatively low altitude. Altitudes below 500-600 kilometers are in general avoided, though, because of the significant air-drag at such low altitudes making frequent orbit reboost maneuvres necessary. The Earth observation satellites ERS-1, ERS-2 and Envisat of European Space Agency as well as the MetOp spacecraft of EUMETSAT are all operated at altitudes of about 800 km. The Proba-1, Proba-2 and SMOS spacecraft of European Space Agency are observing the Earth from an altitude of about 700 km. The Earth observation satellites of UAE, DubaiSat-1 & DubaiSat-2 are also placed in Low Earth Orbits (LEO) orbits and providing satellite imagery of various parts of the Earth.

To get (nearly) global coverage with a low orbit it must be a polar orbit or nearly so. A low orbit will have an orbital period of roughly 100 minutes and the Earth will rotate around its polar axis with about 25 deg between successive orbits, with the result that the ground track is shifted towards west with these 25 deg in longitude. Most are in sun-synchronous orbits.

Spacecraft carrying instruments for which an altitude of 36000 km is suitable sometimes use a geostationary orbit. Such an orbit allows uninterrupted coverage of more than 1/3 of the Earth. Three geostationary spacecraft at longitudes separated with 120 deg can cover the whole Earth except the extreme polar regions. This type of orbit is mainly used for meteorological satellites.

## Weather

A weather satellite is a type of satellite that is primarily used to monitor the weather and climate of the Earth. These meteorological satellites, however, see more than clouds and cloud systems. City lights, fires, effects of pollution, auroras, sand and dust storms, snow cover, ice mapping, boundaries of ocean currents, energy flows, etc., are other types of environmental information collected using weather satellites.

GOES-8, a United States weather satellite.

Weather satellite images helped in monitoring the volcanic ash cloud from Mount St. Helens and activity from other volcanoes such as Mount Etna. Smoke from fires in the western United States such as Colorado and Utah have also been monitored.

The El Niño Southern Oscillation and its effects on weather are monitored daily from satellite images. The Antarctic ozone hole is mapped from weather satellite data. Collectively, weather satellites flown by the U.S., Europe, India, China, Russia, and Japan provide nearly continuous observations for a global weather watch, used via visible light and infrared rays of the electromagnetic spectrum.

## Environmental Monitoring

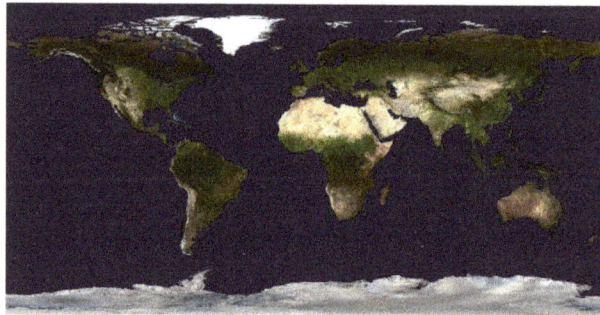

A composite satellite image of the earth, showing its entire surface in Plate carrée projection.

Other environmental satellites can assist environmental monitoring by detecting changes in the Earth's vegetation, atmospheric trace gas content, sea state, ocean color, and ice fields. By monitoring vegetation changes over time, droughts can be monitored by comparing the current vegetation state to its long term average. For example, the 2002 oil spill off the northwest coast of Spain was watched carefully by the European ENVISAT, which, though not a weather satellite, flies an instrument (ASAR) which can see changes in the sea surface. Anthropogenic emissions can be monitored by evaluating data of tropospheric $NO_2$ and $SO_2$.

These types of satellites are almost always in Sun synchronous and "frozen" orbits. The Sun synchronous orbit is in general sufficiently close to polar to get the desired global coverage while the relatively constant geometry to the Sun mostly is an advantage for the instruments. The "frozen" orbit is selected as this is the closest to a circular orbit that is possible in the gravitational field of the Earth

## Mapping

Terrain can be mapped from space with the use of satellites, such as Radarsat-1 and TerraSAR-X.

## Weather Satellite

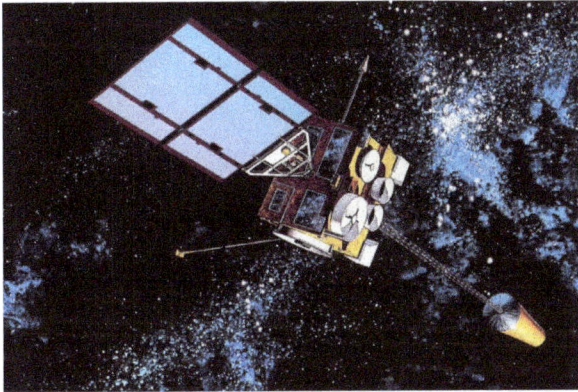

GOES-8, a United States weather satellite of the Meteorological-satellite service.

The weather satellite is a type of satellite that is primarily used to monitor the weather and climate of the Earth. Satellites can be polar orbiting, covering the entire Earth asynchronously, or geostationary, hovering over the same spot on the equator.

Meteorological satellites see more than clouds and cloud systems. City lights, fires, effects of pollution, auroras, sand and dust storms, snow cover, ice mapping, boundaries of ocean currents, energy flows, etc. Other types of environmental information are collected using weather satellites. Weather satellite images helped in monitoring the volcanic ash cloud from Mount St. Helens and activity from other volcanoes such as Mount Etna. Smoke from fires in the western United States such as Colorado and Utah have also been monitored.

Other environmental satellites can detect changes in the Earth's vegetation, sea state, ocean color, and ice fields. For example, the 2002 Prestige oil spill off the northwest coast of Spain was watched carefully by the European ENVISAT, which, though not a weather satellite, flies an instrument (ASAR) which can see changes in the sea surface.

El Niño and its effects on weather are monitored daily from satellite images. The Antarctic ozone hole is mapped from weather satellite data. Collectively, weather satellites flown by the U.S., Europe, India, China, Russia, and Japan provide nearly continuous observations for a global weather watch.

## History

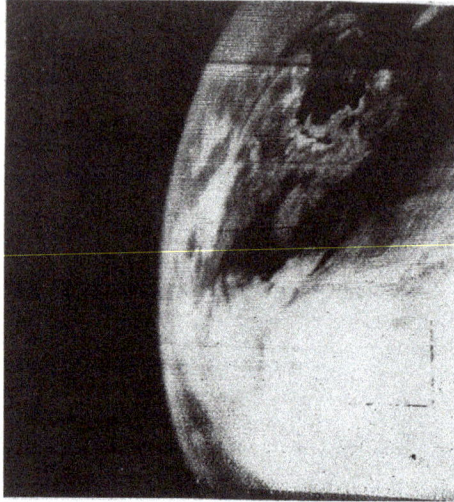

The first television image of Earth from space from the TIROS-1 weather satellite.

As early as 1946, the idea of cameras in orbit to observe the weather was being developed. This was due to sparse data observation coverage and the expense of using cloud cameras on rockets. By 1958, the early prototypes for TIROS and Vanguard (developed by the Army Signal Corps) were created. The first weather satellite, Vanguard 2, was launched on February 17, 1959. It was designed to measure cloud cover and resistance, but a poor axis of rotation and its elliptical orbit kept it from collecting a notable amount of useful data. The Explorer VI and VII satellites also contained weather-related experiments.

The first weather satellite to be considered a success was TIROS-1, launched by NASA on April 1, 1960. TIROS operated for 78 days and proved to be much more successful than Vanguard 2. TIROS paved the way for the Nimbus program, whose technology and findings are the heritage of most of the Earth-observing satellites NASA and NOAA have launched since then. Beginning with the Nimbus 3 satellite in 1969, temperature information through the tropospheric column began to be retrieved by satellites from the eastern Atlantic and most of the Pacific ocean, which led to significant improvements to weather forecasts.

The ESSA and NOAA polar orbiting satellites followed suit from the late 1960s onward. Geostationary satellites followed, beginning with the ATS and SMS series in the late 1960s and early 1970s, then continuing with the GOES series from the 1970s onward. Polar orbiting satellites such as QuikScat and TRMM began to relay wind information near the ocean's surface starting in the late 1970s, with micro-wave imagery which resembled radar displays, which significantly improved the di-agnoses of tropical cyclone strength, intensification, and location during the 2000s and 2010s.

## Observation

Observation is typically made via different 'channels' of the Electromagnetic spectrum, in particular, the Visible and Infrared portions.

Some of these channels include

- *Visible and Near Infrared:* 0.6 μm – 1.6 μm – For recording cloud cover during the day

- *Infrared:* 3.9 μm – 7.3 μm (Water Vapor), 8.7 μm, – 13.4 μm (Thermal imaging)

## Visible Spectrum

Visible-light images from weather satellites during local daylight hours are easy to in-terpret even by the average person; clouds, cloud systems such as fronts and tropical storms, lakes, forests, mountains, snow ice, fires, and pollution such as smoke, smog, dust and haze are readily apparent. Even wind can be determined by cloud patterns, alignments and movement from successive photos.

## Infrared Spectrum

The thermal or infrared images recorded by sensors called scanning radiometers enable a trained analyst to determine cloud heights and types, to calculate land and surface water temperatures, and to locate ocean surface features. Infrared satel-lite imagery can be used effectively for tropical cyclones with a visible eye pattern, using the Dvorak technique, where the difference between the temperature of the warm eye and the surrounding cold cloud tops can be used to determine its inten-sity (colder cloud tops generally indicate a more intense storm). Infrared pictures depict ocean eddies or vortices and map currents such as the Gulf Stream which are valuable to the shipping industry. Fishermen and farmers are interested in knowing land and water temperatures to protect their crops against frost or increase their catch from the sea. Even El Niño phenomena can be spotted. Using color-digitized techniques, the gray shaded thermal images can be converted to color for easier identification of desired information.

## Types

Each meteorological satellite is designed to use one of two different classes of orbit: geostationary and polar orbiting.

The geostationary Himawari 8 satellite's first true-color composite PNG image

## Geostationary

Geostationary weather satellites orbit the Earth above the equator at altitudes of 35,880 km (22,300 miles). Because of this orbit, they remain stationary with respect to the rotating Earth and thus can record or transmit images of the entire hemisphere below continuously with their visible-light and infrared sensors. The news media use the geostationary photos in their daily weather presentation as single images or made into movie loops. These are also available on the city forecast pages of noaa.gov (example Dallas, TX).

Several geostationary meteorological spacecraft are in operation. The United States has three in operation; GOES-12, GOES-13, and GOES-15. GOES-12, previously designated GOES-East and now used for South America, is located at 60 degrees west. GOES-13 took over the role of GOES-East on April 14, 2010 and is located at 75 degrees west. GOES-11 was GOES-West over the eastern Pacific Ocean until it was decommissioned December 2011 and replaced by GOES-15. Russia's new-generation weather satellite Elektro-L No.1 operates at 76°E over the Indian Ocean. The Japanese have the MTSAT-2 located over the mid Pacific at 145°E and the Himawari 8 at 140°E. The Europeans have four in operation, Meteosat-8 (3.5°W) and Meteosat-9 (0°) over the Atlantic Ocean and have Meteosat-6 (63°E) and Meteosat-7 (57.5°E) over the Indian Ocean. China currently has three Fengyun (风云) geostationary satellites (FY-2E at 86.5°E, FY-2F at 123.5°E, and FY-2G at 105°E) operated. India also operates geostationary satellites called INSAT which carry instruments for meteorological purposes.

## Polar Orbiting

Polar orbiting weather satellites circle the Earth at a typical altitude of 850 km (530 miles) in a north to south (or vice versa) path, passing over the poles in their continuous flight. Polar satellites are in sun-synchronous orbits, which means they are able to observe any place on Earth and will view every location twice each day with the same general lighting conditions due to the near-constant local solar time. Polar orbiting weather satellites offer a much better resolution than their geostationary counterparts due their closeness to the Earth.

Computer controlled motorized parabolic dish antenna for tracking LEO weather satellites.

The United States has the NOAA series of polar orbiting meteorological satellites, presently NOAA 17 and NOAA 18 as primary spacecraft, NOAA 15 and NOAA 16 as secondary spacecraft, NOAA 14 in standby, and NOAA 12. Europe has the Metop-A and Metop-B satellites operated by EUMETSAT. Russia has the Meteor and RESURS series of satellites. China has FY-3A, 3B and 3C. India has polar orbiting satellites as well.

## DMSP

Turnstile antenna for reception of 137 MHz LEO weather satellite transmissions

The United States Department of Defense's Meteorological Satellite (DMSP) can "see" the best of all weather vehicles with its ability to detect objects almost as 'small' as a huge oil tanker. In addition, of all the weather satellites in orbit, only DMSP can "see" at night in the visual. Some of the most spectacular photos have been recorded by the night visual sensor; city lights, volcanoes, fires, lightning, meteors, oil field burn-offs, as well as the Aurora Borealis and Aurora Australis have been captured by this 450-mile-high space vehicle's low moonlight sensor.

At the same time, energy use and city growth can be monitored since both major and even minor cities, as well as highway lights, are conspicuous. This informs astronomers of light pollution. The New York City Blackout of 1977 was captured by one of the night orbiter DMSP space vehicles.

In addition to monitoring city lights, these photos are a life saving asset in the detection and monitoring of fires. Not only do the satellites see the fires visually day and night, but the thermal and infrared scanners on board these weather satellites detect potential fire sources below the surface of the Earth where smoldering occurs. Once the fire is detected, the same weather satellites provide vital information about wind that could fan or spread the fires. These same cloud photos from space tell the firefighter when it will rain.

Some of the most dramatic photos showed the 600 Kuwaiti oil fires that the fleeing Army of Iraq started on February 23, 1991. The night photos showed huge flashes, far outstripping the glow of large populated areas. The fires consumed millions of gallons of oil; the last was doused on November 6, 1991.

## Uses

Snowfield monitoring, especially in the Sierra Nevada, can be helpful to the hydrologist keeping track of available snowpack for runoff vital to the watersheds of the western United States. This information is gleaned from existing satellites of all agencies of the U.S. government (in addition to local, on-the-ground measurements). Ice floes, packs and bergs can also be located and tracked from weather space craft.

Even pollution whether it is nature-made or man-made can be pinpointed. The visual and infrared photos show effects of pollution from their respective areas over the entire earth. Aircraft and rocket pollution, as well as condensation trails, can also be spotted. The ocean current and low level wind information gleaned from the space photos can help predict oceanic oil spill coverage and movement. Almost every summer, sand and dust from the Sahara Desert in Africa drifts across the equatorial regions of the Atlantic Ocean. GOES-EAST photos enable meteorologists to observe, track and forecast this sand cloud. In addition to reducing visibilities and causing respiratory problems, sand clouds suppress hurricane formation by modifying the solar radiation balance of the tropics. Other dust storms in Asia and mainland China are common and easy to spot and monitor, with recent examples of dust moving across the Pacific Ocean and reaching North America.

In remote areas of the world with few local observers, fires could rage out of control for days or even weeks and consume millions of acres before authorities are alerted. Weather satellites can be a tremendous asset in such situations. Nighttime photos also show the burn-off in gas and oil fields. Atmospheric temperature and moisture profiles have been taken by weather satellites since 1969.

## Reconnaissance Satellite

U.S. Lacrosse radar spy satellite under construction

A model of a German SAR-Lupe reconnaissance satellite inside a Cosmos-3M rocket.

A reconnaissance satellite (commonly, although unofficially, referred to as a spy satellite) is an Earth observation satellite or communications satellite deployed for military or intelligence applications.

The first generation type (i.e., Corona  and Zenit) took photographs, then ejected canisters of photographic film which would descend to earth. Corona capsules were retrieved in mid-air as they floated down on parachutes. Later, spacecraft had digital imaging systems and downloaded the images via encrypted radio links.

In the United States, most information available is on programs that existed up to 1972, as this information has been declassified due to its age. Some information about programs prior to that time is still classified, and a small amount of information is available on subsequent missions.

A few up-to-date reconnaissance satellite images have been declassified on occasion, or leaked, as in the case of KH-11 photographs which were sent to *Jane's Defence Weekly* in 1984.

## History

On 16 March 1955, the United States Air Force officially ordered the development of an advanced reconnaissance satellite to provide continuous surveillance of 'preselected areas of the earth' in order 'to determine the status of a potential enemy's war-making capability'.

## Types

There are several major types of reconnaissance satellite.

Missile Early warning

> Provides warning of an attack by detecting ballistic missile launches. Earliest known are Missile Defense Alarm System.

Nuclear explosion detection

> Identifies and characterizes nuclear explosions in space. Vela (satellite) is the earliest known.

Photo surveillance

> Provides imaging of earth from space. Images can be a survey or close-look telephoto. Corona (satellite) is the earliest known. Spectral imaging is commonplace.

Electronic-reconnaissance

> Signals intelligence, intercepts stray radio waves. Samos-F is the earliest known.

Radar imaging

> Most space-based radars use synthetic aperture radar. Can be used at night or through cloud cover. Earliest known are US-A series.

## Missions

Examples of reconnaissance satellite missions:

- High resolution photography (IMINT)

- Measurement and Signature Intelligence (MASINT)

- Communications eavesdropping (SIGINT)

- Covert communications

- Monitoring of nuclear test ban compliance

- Detection of missile launches

On 28 August 2013, it was thought that "a $1-billion high-powered spy satellite capable of snapping pictures detailed enough to distinguish the make and model of an automobile hundreds of miles below" was launched from California's Vandenberg Air Force Base using a Delta IV Heavy launcher, America's highest-payload space launch vehicle.

On 17 February 2014, a Russian Kosmos-1220 originally launched in 1980 and used for naval missile targeting until 1982, made an uncontrolled atmospheric entry.

## Benefits

Reconnaissance satellites have been used to enforce human rights, through the Satellite Sentinel Project, which monitors atrocities in Sudan and South Sudan.

During his 1980 State of the Union Address, President Jimmy Carter explained how all of humanity benefited from the presence of American spy satellites:

...photo-reconnaissance satellites, for example, are enormously important in stabilizing world affairs and thereby make a significant contribution to the security of all nations.

Additionally, companies such as GeoEye and DigitalGlobe have provided commercial satellite imagery in support of natural disaster response and humanitarian missions.

During the 1950s, a Soviet hoax had led to American fears of a bomber gap. In 1968, after gaining satellite photography, the United States' intelligence agencies were able to state with certainty that "No new ICBM complexes have been established in the USSR during the past year." President Lyndon B. Johnson told a gathering in 1967:

I wouldn't want to be quoted on this ... We've spent $35 or $40 billion on the space program. And if nothing else had come out of it except the knowledge that we gained from space photography, it would be worth ten times what the whole program has cost. Because tonight we know how many missiles the enemy has and, it turned out, our guesses were way off. We were doing things we didn't need to do. We were building things we didn't need to build. We were harboring fears we didn't need to harbor.

## In Fiction

Spy satellites are commonly seen in spy fiction and military fiction. Some works of fiction that focus specifically on spy satellites include:

- *The OMAC Project*

- *Enemy of the State (film)*

- *Body of Lies (film)*

- *Ice Station Zebra*

- *Karlsson-on-the-Roof is Sneaking Around Again*

## Biosatellite

A biosatellite is a satellite designed to carry life in space. The first satellite carrying an animal (a dog, "Laika") was Soviet Sputnik 2 at November 3, 1957. On August 20, 1960 Soviet Sputnik 5 first time recovered animals (dogs) from orbit to Earth. NASA launched three satellites specifically named Biosatellite (1, 2 & 3) between 1966 and 1969.

NASA's Biosatellite program was a series of three satellites to assess the effects of spaceflight, especially radiation and weightlessness, on living organisms. Each was designed to reenter and be recovered at the end of its mission.

The first two Biosatellites carried specimens of fruit flies, frog eggs, bacteria, and wheat seedlings; the third carried a monkey. Biosatellite 1 was not recovered because of the failure of a retrorocket to ignite. However, Biosatellite-2 successfully deorbited and was recovered in midair by the United States Air Force. Its 13 experiments, exposed to microgravity during a 45-hour orbital flight, provided the first data about basic biological processes in space. Biosatellite 3 carried a 6-kg male pig-tailed monkey, called Bonnie, with the object of investigating the effect of spaceflight on brain states, behavioral performance, cardiovascular status, fluid and electrolyte balance, and metabolic state. Scheduled to remain in orbit for 30 days, the mission was terminated after only 8.8 days because of the subject's deteriorating health. Despite the seeming failure of the mission's scientific agenda, Biosatellite 3 was influential in shaping the life sciences flight experiment program, highlighting the need for centralized management, realistic goals, and adequate preflight experiment verification.

The most famous biosatellites include:

- Biosatellite program launched by NASA between 1966 and 1969.

- Bion space program

- The Mars Gravity Biosatellite

- Orbiting Frog Otolith (OFO-A)

## References

- Martin, Donald; Anderson, Paul; Bartamian, Lucy (March 16, 2007). "Communications Satellites" (5th ed.). AIAA. ISBN 978-1884989193.

- Janice Hill (1991). Weather From Above: America's Meteorological Satellites. Smithsonian Institution. pp. 4–7. ISBN 0-87474-394-X.

- Erickson, Mark. Into the Unknown Together - The DOD, NASA, and Early Spaceflight (PDF). ISBN 1-58566-140-6.

- "Arthur C. Clarke, inventor of satellite, visionary in technology, dead at 90". Engadget.com. 2008-03-18. Retrieved 2016-02-10.

- "Communication: How the rover can communicate through Mars-orbiting spacecraft". Jet Propulsion Laboratory. Retrieved 21 January 2016.

- "Indian GSLV successfully lofts GSAT-14 satellite". NASA Space Flight. 4 January 2014. Retrieved 16 January 2014.

- Hennigan, W.J. (27 August 2013). "Monster rocket to blast off from Pacific coast, rattle Southland". Los Angeles Times. Retrieved 16 February 2014.

- Melissa Goldin (2014-02-17). "Fragments of Soviet-Era Satellite Burn Up in Earth's Atmosphere". Mashable. Retrieved 2014-02-17.

- "The State of the Union Annual Message to the Congress". 1980 State of the Union Address. The American Presidency Project. Retrieved 11 April 2014.

- "Commercial Satellite Imagery Companies Partner with the U.S. Geological Survey in Support of the International Charter "Space and Major Disasters"". USGS Newsroon. United States Geological Survey. Retrieved 4 April 2014.

- Wright, Michael; Herron, Caroline Rand (8 December 1985). "Two Years for Morison". New York Times. Retrieved 16 February 2014.

- India's first 'military' satellite GSAT-7 put into earth's orbit. NDTV.com (2013-09-04). Retrieved on 2013-09-18.

- "GOES 13 Spacecraft Status Summary". NOAA Satellite and Information Service. Retrieved February 15, 2012.

- Ann K. Cook (July 1969). "The Breakthrough Team" (PDF). ESSA World. Environmental Satellite Services Administration: 28–31. Retrieved 2012-04-21.

- "GOES 12 Spacecraft Status Summary". NOAA Satellite and Information Service. Retrieved December 13, 2010.

- "Military Satellite Communications Fundamentals | The Aerospace Corporation". Aerospace. 2010-04-01. Retrieved 2016-02-10.

# Geodesy: Significant Aspects

Geodesy or geodetic engineering is the three-dimensional representation of the Earth's surface by surveying specific regions. It helps in positioning within the temporally varying gravitational field. It is also vital in cartography. This section gives a comprehensive understanding of geodesy to the readers.

## Geodesy

Geodesy also known as geodetics, geodetic engineering or geodetics engineering — a branch of applied mathematics and earth sciences, is the scientific discipline that deals with the measurement and representation of the Earth (or any planet), including its gravitational field, in a three-dimensional time-varying space. Geodesists also study geodynamical phenomena such as crustal motion, tides, and polar motion. For this they design global and national control networks, using space and terrestrial techniques while relying on datums and coordinate systems.

An old geodetic pillar (1855) at Ostend, Belgium

A Munich archive with lithography plates of maps of Bavaria

## Definition

Geodesy is primarily concerned with positioning within the temporally varying gravity field. Geodesy in the German-speaking world is divided into "higher geodesy" ("Erdmessung" or "höhere Geodäsie"), which is concerned with measuring the Earth on the global scale, and "practical geodesy" or "engineering geodesy" ("Ingenieurgeodäsie"), which is concerned with measuring specific parts or regions of the Earth, and which includes surveying. Such "geodetic" operations are also applied to other astronomical bodies in the solar system. It is also the science of measuring and understanding the earth's geometric shape, orientation in space and gravity field.

The shape of the Earth is to a large extent the result of its rotation, which causes its equatorial bulge, and the competition of geological processes such as the collision of plates and of volcanism, resisted by the Earth's gravity field. This applies to the solid surface, the liquid surface (dynamic sea surface topography) and the Earth's atmosphere. For this reason, the study of the Earth's gravity field is called physical geodesy by some.

## History

### Geoid and Reference Ellipsoid

The geoid is essentially the figure of the Earth abstracted from its topographical features. It is an idealized equilibrium surface of sea water, the mean sea level surface in the absence of currents, air pressure variations etc. and continued under the continental masses. The geoid, unlike the reference ellipsoid, is irregular and too complicated to serve as the computational surface on which to solve geometrical problems like point positioning. The geometrical separation between the geoid and the reference ellipsoid is called the geoidal undulation. It varies globally between ±110 m, when referred to the GRS 80 ellipsoid.

A reference ellipsoid, customarily chosen to be the same size (volume) as the geoid, is described by its semi-major axis (equatorial radius) $a$ and flattening $f$. The quantity $f = a - b/a$, where $b$ is the semi-minor axis (polar radius), is a purely geometrical one. The mechanical ellipticity of the Earth (dynamical flattening, symbol $J_2$) can be determined to high precision by observation of satellite orbit perturbations. Its relationship with the geometrical flattening is indirect. The relationship depends on the internal density distribution, or, in simplest terms, the degree of central concentration of mass.

The 1980 Geodetic Reference System (GRS 80) posited a 6,378,137 m semi-major axis and a 1:298.257 flattening. This system was adopted at the XVII General Assembly of the International Union of Geodesy and Geophysics (IUGG). It is essentially the basis for geodetic positioning by the Global Positioning System and is thus also in widespread use outside the geodetic community.

The numerous other systems which have been used by diverse countries for their maps and charts are gradually dropping out of use as more and more countries move to global, geocentric reference systems using the GRS 80 reference ellipsoid.

## Coordinate Systems in Space

The locations of points in three-dimensional space are most conveniently described by three cartesian or rectangular coordinates, $X$, $Y$ and $Z$. Since the advent of satellite positioning, such coordinate systems are typically geocentric: the $Z$-axis is aligned with the Earth's (conventional or instantaneous) rotation axis.

Prior to the era of satellite geodesy, the coordinate systems associated with a geodetic datum attempted to be geocentric, but their origins differed from the geocentre by hundreds of metres, due to regional deviations in the direction of the plumbline (vertical). These regional geodetic data, such as ED 50 (European Datum 1950) or NAD 27 (North American Datum 1927) have ellipsoids associated with them that are regional 'best fits' to the geoids within their areas of validity, minimising the deflections of the vertical over these areas.

It is only because GPS satellites orbit about the geocentre, that this point becomes naturally the origin of a coordinate system defined by satellite geodetic means, as the satellite positions in space are themselves computed in such a system.

Geocentric coordinate systems used in geodesy can be divided naturally into two classes:

1. Inertial reference systems, where the coordinate axes retain their orientation relative to the fixed stars, or equivalently, to the rotation axes of ideal gyroscopes; the $X$-axis points to the vernal equinox

2. Co-rotating, also ECEF ("Earth Centred, Earth Fixed"), where the axes are attached to the solid body of the Earth. The $X$-axis lies within the Greenwich observatory's meridian plane.

The coordinate transformation between these two systems is described to good approximation by (apparent) sidereal time, which takes into account variations in the Earth's axial rotation (length-of-day variations). A more accurate description also takes polar motion into account, a phenomenon closely monitored by geodesists.

## Coordinate Systems in the Plane

In surveying and mapping, important fields of application of geodesy, two general types of coordinate systems are used in the plane:

1. Plano-polar, in which points in a plane are defined by a distance $s$ from a specified point along a ray having a specified direction $\alpha$ with respect to a base line or axis;

2. Rectangular, points are defined by distances from two perpendicular axes called $x$ and $y$. It is geodetic practice—contrary to the mathematical convention—to let the $x$-axis point to the north and the $y$-axis to the east.

Rectangular coordinates in the plane can be used intuitively with respect to one's current location, in which case the $x$-axis will point to the local north. More formally, such coordinates can be obtained from three-dimensional coordinates using the artifice of a map projection. It is *not* possible to map the curved surface of the Earth onto a flat map surface without deformation. The compromise most often chosen—called a conformal projection—preserves angles and length ratios, so that small circles are mapped as small circles and small squares as squares.

An example of such a projection is UTM (Universal Transverse Mercator). Within the map plane, we have rectangular coordinates $x$ and $y$. In this case the north direction used for reference is the *map* north, not the *local* north. The difference between the two is called meridian convergence.

It is easy enough to "translate" between polar and rectangular coordinates in the plane: let, as above, direction and distance be $\alpha$ and $s$ respectively, then we have

$$x = s \cos \alpha$$

$$y = s \sin \alpha$$

The reverse transformation is given by:

$$s = \sqrt{x^2 + y^2}$$

$$\alpha = \arctan \frac{y}{x}.$$

## Heights

In geodesy, point or terrain *heights* are "above sea level", an irregular, physically defined surface. Therefore, a height should ideally *not* be referred to as a coordinate. It is more like a physical quantity, and though it can be tempting to treat height as the vertical coordinate $z$, in addition to the horizontal coordinates $x$ and $y$, and though this actually is a good approximation of physical reality in small areas, it quickly becomes invalid for regional considerations.

Heights come in the following variants:

1. Orthometric heights

2. Normal heights

3. Geopotential heights

Each has its advantages and disadvantages. Both orthometric and normal heights are heights in metres above sea level, whereas geopotential numbers are measures of potential energy (unit: $m^2\ s^{-2}$) and not metric. Orthometric and normal heights differ in the precise way in which mean sea level is conceptually continued under the continental masses. The reference surface for orthometric heights is the geoid, an equipotential surface approximating mean sea level.

*None* of these heights is in any way related to geodetic or ellipsoidial heights, which express the height of a point above the reference ellipsoid. Satellite positioning receivers typically provide ellipsoidal heights, unless they are fitted with special conversion software based on a model of the geoid.

## Geodetic Data

Because geodetic point coordinates (and heights) are always obtained in a system that has been constructed itself using real observations, geodesists introduce the concept of a *geodetic datum*: a physical realization of a coordinate system used for describing point locations. The realization is the result of *choosing* conventional coordinate values for one or more *datum points*.

In the case of height data, it suffices to choose *one* datum point: the reference bench mark, typically a tide gauge at the shore. Thus we have vertical data like the NAP (Normaal Amsterdams Peil), the North American Vertical Datum 1988 (NAVD 88), the Kronstadt datum, the Trieste datum, and so on.

In case of plane or spatial coordinates, we typically need several datum points. A regional, ellipsoidal datum like ED 50 can be fixed by prescribing the undulation of the geoid and the deflection of the vertical in *one* datum point, in this case the Helmert Tower in Potsdam. However, an overdetermined ensemble of datum points can also be used.

Changing the coordinates of a point set referring to one datum, so to make them refer to another datum, is called a *datum transformation*. In the case of vertical data, this consists of simply adding a constant shift to all height values. In the case of plane or spatial coordinates, datum transformation takes the form of a similarity or *Helmert transformation*, consisting of a rotation and scaling operation in addition to a simple translation. In the plane, a Helmert transformation has four parameters; in space, seven.

## A Note on Terminology

In the abstract, a coordinate system as used in mathematics and geodesy is, e.g., in ISO terminology, referred to as a *coordinate system*. International geodetic organizations like the IERS (International Earth Rotation and Reference Systems Service) speak of a *reference system*.

When these coordinates are realized by choosing datum points and fixing a geodetic datum, ISO uses the terminology *coordinate reference system*, while IERS speaks of a *reference frame*. A datum transformation again is referred to by ISO as a *coordinate transformation*. (ISO 19111: Spatial referencing by coordinates).

## Point Positioning

Geodetic Control Mark (example of a deep benchmark)

Point positioning is the determination of the coordinates of a point on land, at sea, or in space with respect to a coordinate system. Point position is solved by computation from measurements linking the known positions of terrestrial or extraterrestrial points with the unknown terrestrial position. This may involve transformations between or among astronomical and terrestrial coordinate systems.

The known points used for point positioning can be triangulation points of a higher order network, or GPS satellites.

Traditionally, a hierarchy of networks has been built to allow point positioning within a country. Highest in the hierarchy were triangulation networks. These were densified into networks of traverses (polygons), into which local mapping surveying measurements, usually with measuring tape, corner prism and the familiar red and white poles, are tied.

Nowadays all but special measurements (e.g., underground or high precision engineering measurements) are performed with GPS. The higher order networks are measured with static GPS, using differential measurement to determine vectors between terrestrial points. These vectors are then adjusted in traditional network fashion. A global polyhedron of permanently operating GPS stations under the auspices of the IERS is used to define a single global, geocentric reference frame which serves as the "zero order" global reference to which national measurements are attached.

For surveying mappings, frequently Real Time Kinematic GPS is employed, tying in the unknown points with known terrestrial points close by in real time.

One purpose of point positioning is the provision of known points for mapping measurements, also known as (horizontal and vertical) control. In every country, thousands of such known points exist and are normally documented by the national mapping agencies. Surveyors involved in real estate and insurance will use these to tie their local measurements to.

## Geodetic Problems

In geometric geodesy, two standard problems exist:

### First (Direct) Geodetic Problem

Given a point (in terms of its coordinates) and the direction (azimuth) and distance from that point to a second point, determine (the coordinates of) that second point.

### Second (Inverse) Geodetic Problem

Given two points, determine the azimuth and length of the line (straight line, arc or geodesic) that connects them.

In the case of plane geometry (valid for small areas on the Earth's surface) the solutions to both problems reduce to simple trigonometry. On the sphere, the solution is significantly more complex, e.g., in the inverse problem the azimuths will differ between the two end points of the connecting great circle, arc, i.e. the geodesic.

On the ellipsoid of revolution, geodesics may be written in terms of elliptic integrals, which are usually evaluated in terms of a series expansion.

In the general case, the solution is called the geodesic for the surface considered. The differential equations for the geodesic can be solved numerically.

## Geodetic Observational Concepts

Here we define some basic observational concepts, like angles and coordinates, defined in geodesy (and astronomy as well), mostly from the viewpoint of the local observer.

- The *plumbline* or *vertical* is the direction of local gravity, or the line that results by following it.

- The *zenith* is the point on the celestial sphere where the direction of the gravity vector in a point, extended upwards, intersects it. More correct is to call it a *direction* rather than a point.

- The *nadir* is the opposite point (or rather, direction), where the direction of gravity extended downward intersects the (invisible) celestial sphere.

- The celestial *horizon* is a plane perpendicular to a point's gravity vector.

- *Azimuth* is the direction angle within the plane of the horizon, typically counted clockwise from the north (in geodesy and astronomy) or south (in France).

- *Elevation* is the angular height of an object above the horizon, Alternatively zenith distance, being equal to 90 degrees minus elevation.

- *Local topocentric coordinates* are azimuth (direction angle within the plane of the horizon) and elevation angle (or zenith angle) and distance.

- The north *celestial pole* is the extension of the Earth's (precessing and nutating) instantaneous spin axis extended Northward to intersect the celestial sphere. (Similarly for the south celestial pole.)

- The *celestial equator* is the intersection of the (instantaneous) Earth equatorial plane with the celestial sphere.

- A *meridian plane* is any plane perpendicular to the celestial equator and containing the celestial poles.

- The *local meridian* is the plane containing the direction to the zenith and the direction to the celestial pole.

## Geodetic Measurements

The level is used for determining height differences and height reference systems, commonly referred to mean sea level. The traditional spirit level produces these practically most useful heights above sea level directly; the more economical use of GPS instruments for height determination requires precise knowledge of the figure of the geoid, as GPS only gives heights above the GRS80 reference ellipsoid. As geoid knowledge accumulates, one may expect use of GPS heighting to spread.

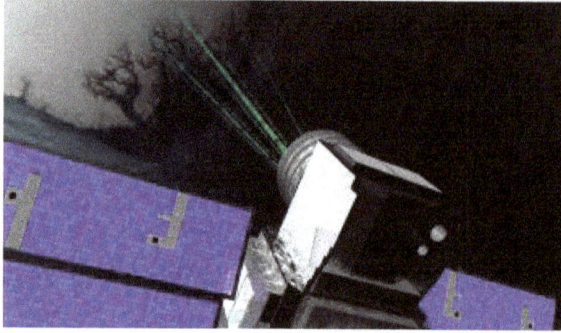

Project manager Stephen Merkowitz talks about his work with NASA's Space Geodesy Project, including a brief overview of the four fundamental techniques of space geodesy: GPS, VLBI, SLR, and DORIS.

The theodolite is used to measure horizontal and vertical angles to target points. These angles are referred to the local vertical. The tacheometer additionally determines, electronically or electro-optically, the distance to target, and is highly automated to even robotic in its operations. The method of free station position is widely used.

For local detail surveys, tacheometers are commonly employed although the old-fashioned rectangular technique using angle prism and steel tape is still an inexpensive alternative. Real-time kinematic (RTK) GPS techniques are used as well. Data collected are tagged and recorded digitally for entry into a Geographic Information System (GIS) database.

Geodetic GPS receivers produce directly three-dimensional coordinates in a geocentric coordinate frame. Such a frame is, e.g., WGS84, or the frames that are regularly produced and published by the International Earth Rotation and Reference Systems Service (IERS).

GPS receivers have almost completely replaced terrestrial instruments for large-scale base network surveys. For Planet-wide geodetic surveys, previously impossible, we can still mention Satellite Laser Ranging (SLR) and Lunar Laser Ranging (LLR) and Very Long Baseline Interferometry (VLBI) techniques. All these techniques also serve to monitor Earth rotation irregularities as well as plate tectonic motions.

Gravity is measured using gravimeters. Basically, there are two kinds of gravimeters. *Absolute* gravimeters, which nowadays can also be used in the field, are based directly on measuring the acceleration of free fall (for example, of a reflecting prism in a vacuum tube). They are used for establishing the vertical geospatial control. Most common *relative* gravimeters are spring based. They are used in gravity surveys over large areas for establishing the figure of the geoid over these areas. Most accurate relative gravimeters are *superconducting* gravimeters, and these are sensitive to one thousandth of one billionth of Earth surface gravity. Twenty-some superconducting gravimeters are used worldwide for studying Earth tides, rotation, interior, and ocean and atmospheric loading, as well as for verifying the Newtonian constant of gravitation.

In the future gravity, and altitude, will be measured by relativistic time dilation measured by strontium optical clocks.

## Units and Measures on the Ellipsoid

Geographical latitude and longitude are stated in the units degree, minute of arc, and second of arc. They are *angles*, not metric measures, and describe the *direction* of the local normal to the reference ellipsoid of revolution. This is *approximately* the same as the direction of the plumbline, i.e., local gravity, which is also the normal to the geoid surface. For this reason, astronomical position determination – measuring the direction of the plumbline by astronomical means – works fairly well provided an ellipsoidal model of the figure of the Earth is used.

One geographical mile, defined as one minute of arc on the equator, equals 1,855.32571922 m. One nautical mile is one minute of astronomical latitude. The radius of curvature of the ellipsoid varies with latitude, being the longest at the pole and the shortest at the equator as is the nautical mile.

A metre was originally defined as the 10-millionth part of the length of a meridian (the target was not quite reached in actual implementation, so that is off by 200 ppm in the current definitions). This means that one kilometre is roughly equal to (1/40,000) * 360 * 60 meridional minutes of arc, which equals 0.54 nautical mile, though this is not exact because the two units are defined on different bases (the international nautical mile is defined as exactly 1,852 m, corresponding to a rounding of 1000/0.54 m to four digits).

## Temporal Change

In geodesy, temporal change can be studied by a variety of techniques. Points on the Earth's surface change their location due to a variety of mechanisms:

- Continental plate motion, plate tectonics

- Episodic motion of tectonic origin, esp. close to fault lines

- Periodic effects due to Earth tides

- Postglacial land uplift due to isostatic adjustment

- Mass variations due to hydrological changes

- Various anthropogenic movements due to, for instance, petroleum or water extraction or reservoir construction.

The science of studying deformations and motions of the Earth's crust and the solid Earth as a whole is called geodynamics. Often, study of the Earth's irregular rotation is also included in its definition.

Techniques for studying geodynamic phenomena on the global scale include:

- satellite positioning by GPS and other such systems,

- Very Long Baseline Interferometry (VLBI)

- satellite and lunar laser ranging

- Regionally and locally, precise levelling,

- precise tacheometers,

- monitoring of gravity change,

- Interferometric synthetic aperture radar (InSAR) using satellite images, etc.

## Geodetic Datum

A geodetic datum or geodetic system is a coordinate system, and a set of reference points, used to locate places on the Earth (or similar objects). An approximate definition of sea level is the datum WGS 84, an ellipsoid, whereas a more accurate definition is Earth Gravitational Model 2008 (EGM2008), using at least 2,159 spherical harmonics. Other datums are defined for other areas or at other times; ED50 was defined in 1950 over Europe and differs from WGS 84 by a few hundred meters depending on where in Europe you look. Mars has no oceans and so no sea level, but at least two martian datums have been used to locate places there.

Datums are used in geodesy, navigation, and surveying by cartographers and satellite navigation systems to translate positions indicated on maps (paper or digital) to their real position on Earth. Each starts with an ellipsoid (stretched sphere), and then defines latitude, longitude and altitude coordinates. One or more locations on the Earth's surface is chosen as an anchor "base-point".

The difference in co-ordinates between datums is commonly referred to as *datum shift*. The datum shift between two particular datums can vary from one place to another within one country or region, and can be anything from zero to hundreds of meters (or several kilometers for some remote islands). The North Pole, South Pole and Equator will be in different positions on different datums, so True North will be slightly different. Different datums use different interpolations for the precise shape and size of the Earth (reference ellipsoids).

Because the Earth is an imperfect ellipsoid, localised datums can give a more accurate representation of the area of coverage than WGS 84. OSGB36, for example, is a better approximation to the geoid covering the British Isles than the global WGS 84 ellipsoid. However, as the benefits of a global system outweigh the greater accuracy, the global WGS 84 datum is becoming increasingly adopted.

Horizontal datums are used for describing a point on the Earth's surface, in latitude and longitude or another coordinate system. Vertical datums measure elevations or depths.

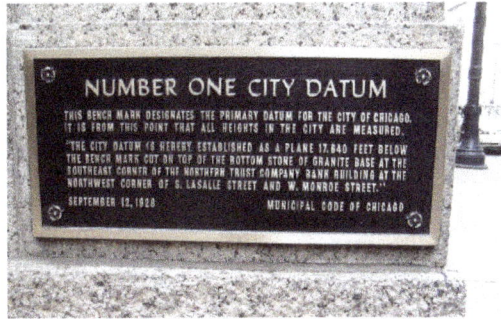

City of Chicago Datum Benchmark

## Definition

In surveying and geodesy, a *datum* is a reference system or an approximation of the Earth's surface against which positional measurements are made for computing locations. Horizontal datums are used for describing a point on the Earth's surface, in latitude and longitude or another coordinate system. Vertical datums are used to measure elevations or underwater depths.

## Horizontal Datum

The horizontal datum is the model used to measure positions on the Earth. A specific point on the Earth can have substantially different coordinates, depending on the datum used to make the measurement. There are hundreds of local horizontal datums around the world, usually referenced to some convenient local reference point. Contemporary datums, based on increasingly accurate measurements of the shape of the Earth, are intended to cover larger areas. The WGS 84 datum, which is almost identical to the NAD83 datum used in North America and the ETRS89 datum used in Europe, is a common standard datum.

For example, in Sydney there is a 200 metres (700 feet) difference between GPS coordinates configured in GDA (based on global standard WGS 84) and AGD (used for most local maps), which is an unacceptably large error for some applications, such as surveying or site location for scuba diving.

## Vertical Datum

A vertical datum is used as a reference point for elevations of surfaces and features on the Earth including terrain, bathymetry, water levels, and man-made structures. Vertical datums are either: tidal, based on sea levels; gravimetric, based on a geoid; or geodetic, based on the same ellipsoid models of the Earth used for computing horizontal datums.

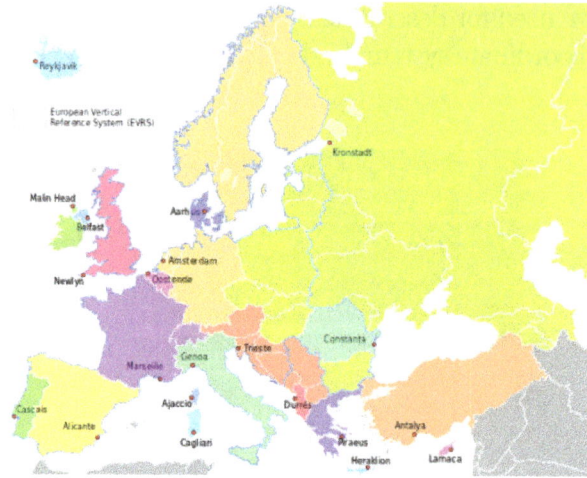

Vertical datums in Europe

In common usage, elevations are often cited in height above sea level, although what "sea level" actually means is a more complex issue than might at first be thought: the height of the sea surface at any one place and time is a result of numerous effects, including waves, wind and currents, atmospheric pressure, tides, topography, and even differences in the strength of gravity due to the presence of mountains etc.

For the purpose of measuring the height of objects on land, the usual datum used is mean sea level (MSL). This is a tidal datum which is described as the arithmetic mean of the hourly water elevation taken over a specific 19 years cycle. This definition averages out tidal highs and lows (caused by the gravitational effects of the sun and the moon) and short term variations. It will not remove the effects of local gravity strength, and so the height of MSL, relative to a geodetic datum, will vary around the world, and even around one country. Countries tend to choose the mean sea level at one specific point to be used as the standard "sea level" for all mapping and surveying in that country. (For example, in Great Britain, the national vertical datum, Ordnance Datum Newlyn, is based on what was mean sea level at Newlyn in Cornwall between 1915 and 1921). However, zero elevation as defined by one country is not the same as zero elevation defined by another (because MSL is not the same everywhere), which is why locally defined vertical datums differ from one another.

A different principle is used when choosing a datum for nautical charts. For safety reasons, a mariner must be able to know the minimum depth of water that could occur at any point. For this reason, depths and tides on a nautical chart are measured relative to chart datum, which is defined to be a level below which tide rarely falls. Exactly how this is chosen depends on the tidal regime in the area being charted and on the policy of the hydrographic office producing the chart in question; a typical definition is Lowest Astronomical Tide (the lowest tide predictable from the effects of gravity), or Mean Lower Low Water (the average lowest tide of each day), although MSL is sometimes used in waters with very low tidal ranges.

Conversely, if a ship is to safely pass under a low bridge or overhead power cable, the mariner must know the minimum clearance between the masthead and the obstruction, which will occur at high tide. Consequently, bridge clearances etc. are given relative to a datum based on high tide, such as Highest Astronomical Tide or Mean High Water Springs.

Sea level does not remain constant throughout geological time, and so tidal datums are less useful when studying very long-term processes. In some situations sea level does not apply at all — for instance for mapping Mars' surface — forcing the use of a different "zero elevation", such as mean radius.

A geodetic vertical datum takes some specific zero point, and computes elevations based on the geodetic model being used, without further reference to sea levels. Usually, the starting reference point is a tide gauge, so at that point the geodetic and tidal datums might match, but due to sea level variations, the two scales may not match elsewhere. An example of a gravity-based geodetic datum is NAVD88, used in North America, which is referenced to a point in Quebec, Canada. Ellipsoid-based datums such as WGS 84, GRS80 or NAD83 use a theoretical surface that may differ significantly from the geoid.

## Geodetic Coordinates

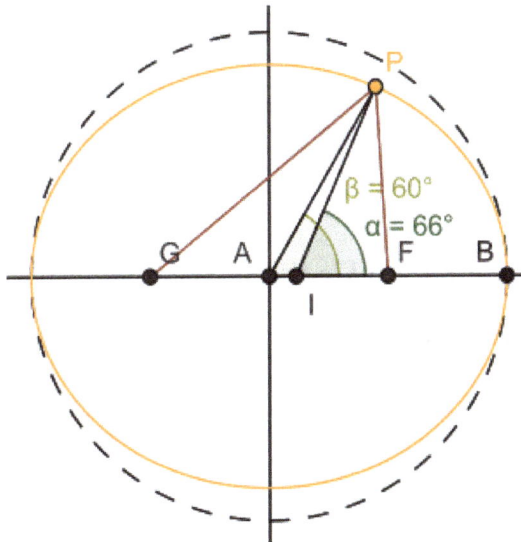

The same position on a spheroid has a different angle for latitude depending on whether the angle is measured from the normal line segment *IP* of the ellipsoid (angle $\alpha$) or the line segment *AP* from the center (angle $\beta$). Note that the "flatness" of the spheroid (orange) in the image is greater than that of the Earth; as a result, the corresponding difference between the "geodetic" and "geocentric" latitudes is also exaggerated.

In geodetic coordinates the Earth's surface is approximated by an ellipsoid and locations near the surface are described in terms of latitude ($\phi$), longitude ($\lambda$) and height ($h$).

## Geodetic Versus Geocentric Latitude

It is important to note that geodetic latitude ( $\phi$ ) (resp. altitude) is different from geo-centric latitude ( $\phi'$ ) (resp. altitude). Geodetic latitude is determined by the angle be-tween the equatorial plane and normal to the ellipsoid, whereas geocentric latitude is determined by the angle between the equatorial plane and line joining the point to the centre of the ellipsoid. Unless otherwise specified latitude is geodetic latitude.

## Earth Reference Ellipsoid

## Defining and Derived Parameters

The ellipsoid is completely parameterised by the semi-major axis $a$ and the flattening $f$.

| Parameter | Symbol |
|---|---|
| Semi-major axis | $a$ |
| Reciprocal of flattening | $1/f$ |

From $a$ and $f$ it is possible to derive the semi-minor axis $b$, first eccentricity $e$ and second eccentricity $e'$ of the ellipsoid

| Parameter | Value |
|---|---|
| semi-minor axis | $b = a(1 - f)$ |
| First eccentricity squared | $e^2 = 1 - b^2/a^2 = 2f - f^2$ |
| Second eccentricity squared | $e'^2 = a^2/b^2 - 1 = f(2 - f)/(1 - f)^2$ |

## Parameters for Some Geodetic Systems

Australian Geodetic Datum 1966 [AGD66] and Australian Geodetic Datum 1984 (AGD84)

AGD66 and AGD84 both use the parameters defined by Australian National Spheroid

Australian National Spheroid (ANS)

| ANS Defining Parameters | | |
|---|---|---|
| Parameter | Notation | Value |
| semi-major axis | a | 6 378 160.000 m |
| Reciprocal of Flattening | $1/f$ | 298.25 |

Geocentric Datum of Australia 1994 (GDA94)

GDA94 uses the parameters defined by GRS80

Geodetic Reference System 1980 (GRS80)

| GRS80 Parameters | | |
|---|---|---|
| **Parameter** | **Notation** | **Value** |
| semi-major axis | a | 6 378 137 m |
| Reciprocal of flattening | $1/f$ | 298.257 222 101 |

see GDA Technical Manual document for more details; the value given above for the flattening is not exact.

World Geodetic System 1984 (WGS 84)

The Global Positioning System (GPS) uses the World Geodetic System 1984 (WGS 84) to determine the location of a point near the surface of the Earth.

| WGS 84 Defining Parameters | | |
|---|---|---|
| **Parameter** | **Notation** | **Value** |
| semi-major axis | a | 6 378 137.0 m |
| Reciprocal of flattening | $1/f$ | 298.257 223 563 |

| WGS 84 derived geometric constants | | |
|---|---|---|
| **Constant** | **Notation** | **Value** |
| Semi-minor axis | $b$ | 6 356 752.3142 m |
| First eccentricity squared | $e^2$ | $6.694\ 379\ 990\ 14 \times 10^{-3}$ |
| Second eccentricity squared | $e'^2$ | $6.739\ 496\ 742\ 28 \times 10^{-3}$ |

see The official World Geodetic System 1984 document for more details.

A more comprehensive list of geodetic systems can be found here

## Conversion Calculations

Datum conversion is the process of converting the coordinates of a point from one datum system to another. Datum conversion may frequently be accompanied by a change of grid projection.

## Reference Datums

A reference datum is a known and constant surface which is used to describe the location of unknown points on the Earth. Since reference datums can have different radii and different center points, a specific point on the Earth can have substantially different coordinates depending on the datum used to make the measurement. There are hundreds of locally developed reference datums around the world, usually referenced to some convenient local reference point. Contemporary datums, based on increasingly accurate measurements of the shape of the Earth, are intended to cover larger areas. The most common reference Datums in use in North America are NAD27, NAD83, and WGS 84.

The North American Datum of 1927 (NAD 27) is "the horizontal control datum for the United States that was defined by a location and azimuth on the Clarke spheroid of 1866, with origin at (the survey station) Meades Ranch (Kansas)." ... The geoidal height at Meades Ranch was assumed to be zero. "Geodetic positions on the North American Datum of 1927 were derived from the (coordinates of and an azimuth at Meades Ranch) through a readjustment of the triangulation of the entire network in which Laplace azimuths were introduced, and the Bowie method was used." (http://www.ngs.noaa.gov/faq.shtml#WhatDatum ) NAD27 is a local referencing system covering North America.

The North American Datum of 1983 (NAD 83) is "The horizontal control datum for the United States, Canada, Mexico, and Central America, based on a geocentric origin and the Geodetic Reference System 1980 (GRS80). "This datum, designated as NAD 83 ...is based on the adjustment of 250,000 points including 600 satellite Doppler stations which constrain the system to a geocentric origin." NAD83 may be considered a local referencing system.

WGS 84 is the World Geodetic System of 1984. It is the reference frame used by the U.S. Department of Defense (DoD) and is defined by the National Geospatial-Intelligence Agency (NGA) (formerly the Defense Mapping Agency, then the National Imagery and Mapping Agency). WGS 84 is used by DoD for all its mapping, charting, surveying, and navigation needs, including its GPS "broadcast" and "precise" orbits. WGS 84 was defined in January 1987 using Doppler satellite surveying techniques. It was used as the reference frame for broadcast GPS Ephemerides (orbits) beginning January 23, 1987. At 0000 GMT January 2, 1994, WGS 84 was upgraded in accuracy using GPS measurements. The formal name then became WGS 84 (G730), since the upgrade date coincided with the start of GPS Week 730. It became the reference frame for broadcast orbits on June 28, 1994. At 0000 GMT September 30, 1996 (the start of GPS Week 873), WGS 84 was redefined again and was more closely aligned with International Earth Rotation Service (IERS) frame ITRF 94. It was then formally called WGS 84 (G873). WGS 84 (G873) was adopted as the reference frame for broadcast orbits on January 29, 1997. Another update brought it to WGS84(G1674).

The WGS 84 datum, within two meters of the NAD83 datum used in North America, is the only world referencing system in place today. WGS 84 is the default standard datum for coordinates stored in recreational and commercial GPS units.

Users of GPS are cautioned that they must always check the datum of the maps they are using. To correctly enter, display, and to store map related map coordinates, the datum of the map must be entered into the GPS map datum field.

- Hong Kong Principal Datum, a vertical datum used in Hong Kong.

# Regional Classification of Satellite Navigation System

This chapter inspects the various satellites used in positioning based on region. The satellites studied include Indian Regional Navigation Satellite System (India), Quasi-Zenith Satellite System (Japan) and DORIS (France). The reader is provided with information on the orbit, positioning and special functions. The aspects elucidated in this chapter are of vital importance, and provide a better understanding of satellite navigation systems.

## Indian Regional Navigation Satellite System

The Indian Regional Navigation Satellite System (IRNSS) with an operational name of NAVIC ("sailor" or "navigator" in Sanskrit, Hindi and many other Indian languages, which also stands for NAVigation with Indian Constellation) is an autonomous regional satellite navigation system that is being set up by India, that will be used to provide accurate real-time positioning and timing services over India and the region extending to 1,500 kilometres (930 mi) around India. The NAVIC system will consist of a constellation of 3 satellites in Geostationary orbit (GEO), 4 satellites in Geosynchronous orbit (GSO), approximately 36,000 kilometres (22,000 mi) altitude above earth surface, and two satellites on the ground as stand-by, in addition to ground stations. The system was developed because access to foreign government-controlled global navigation satellite systems is not guaranteed in hostile situations, as happened to the Indian military in 1999 when it was dependent on the American Global Positioning System (GPS) during the Kargil War. The Indian government approved the project in May 2006.

The constellation of seven NAVIC satellites is already in orbit and the system is expected to be operational from September 2016, after a system check. NAVIC will provide two levels of service, the standard positioning service will be open for civilian use, and a restricted service (an encrypted one) for authorized users (including the military).

### Development

As part of the project, the Indian Space Research Organisation (ISRO) opened a new satellite navigation center within the campus of ISRO Deep Space Network (DSN) at Byalalu, in Karnataka on 28 May 2013. A network of 21 ranging stations located across the country will provide data for the orbital determination of the satellites and monitoring of the navigation signal.

A goal of complete Indian control has been stated, with the space segment, ground segment and user receivers all being built in India. Its location in low latitudes facilitates a coverage with low-inclination satellites. Three satellites will be in geostationary orbit over the Indian Ocean. Missile targeting could be an important military application for the constellation.

The total cost of the project is expected to be ₹1,420 crore (US$211 million), with the cost of the ground segment being ₹300 crore (US$45 million). Each satellites costing ₹150 crore (US$22 million) and the PSLV-XL version rocket costs around ₹130 crore (US$19 million) . The seven rockets would involve an outlay of around ₹910 crore (US$135 million). The NAVIC signal was released for evaluation in September 2014.

## Time-frame

In April 2010, it was reported that India plans to start launching satellites by the end of 2011, at a rate of one satellite every six months. This would have made NAVIC functional by 2015. But the program was delayed, and India also launched 3 new satellites to supplement this.

Seven satellites with the prefix "IRNSS-1" will constitute the space segment of the IRNSS. IRNSS-1A, the first of the seven satellites, was launched on 1 July 2013. IRNSS-1B was launched on 4 April 2014 on board the PSLV-C24 rocket. The satellite has been placed in geosynchronous orbit. IRNSS-1C was launched on 16 October 2014, IRNSS-1D on 28 March 2015, IRNSS-1E on 20 January 2016,IRNSS-1F on 10 March 2016 and IRNSS-1G was launched on 28 April 2016.

## Description

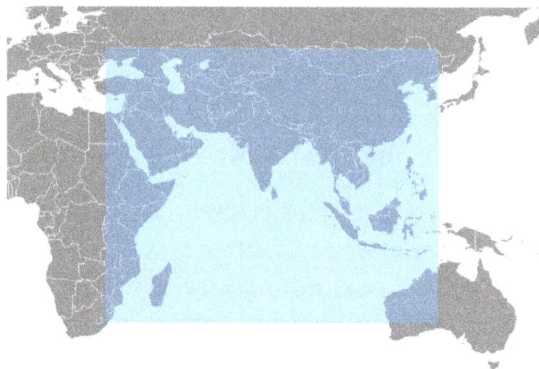

Coverage of the NAVIC.

The system consists of a constellation of seven satellites and a support ground segment. Three of the satellites in the constellation are located in geostationary orbit (GEO) at 32.5° East, 83° East, and 131.5° East longitude. The other four are inclined geosynchronous orbit (GSO). Two of the GSOs cross the equator at 55° East and two at 111.75° East. The four GSO satellites will appear to be moving in the form of an "8".

In addition, various ground-based systems will control, track orbits, check integration and send radio signals to the satellites. The land-based Master Control Center (MCC) will run navigational software.

NAVIC signals will consist of a Standard Positioning Service and a Precision Service. Both will be carried on L5 (1176.45 MHz) and S band (2492.028 MHz). The SPS signal will be modulated by a 1 MHz BPSK signal. The Precision Service will use BOC(5,2). The navigation signals themselves would be transmitted in the S-band frequency (2–4 GHz) and broadcast through a phased array antenna to maintain required coverage and signal strength. The satellites would weigh approximately 1,330 kg and their solar panels generate 1,400 watts. The system is intended to provide an absolute position accuracy of better than 10 meters throughout Indian landmass and better than 20 meters in the Indian Ocean as well as a region extending approximately 1,500 km around India.

## Satellites

### IRNSS-1A

IRNSS-1A was the first out of seven navigational satellite in the Indian Regional Navigation Satellite System (IRNSS) series of satellites to be placed in geosynchronous orbit. It was built at ISRO Satellite Centre, Bangalore, costing ₹125 crore (US$19 million). It has a lift-off mass of 1380 kg, and carries a navigation payload and a C-band ranging transponder, which operates in L5 band (1176.45 MHz) and S band (2492.028 MHz). An optimised I-1K bus structure with a power handling capability of around 1600 watts is used and is designed for a ten-year mission. The satellite was launched on-board PSLV-C22 on 1 July 2013 from the Satish Dhawan Space Centre at Sriharikota.

### IRNSS-1B

Satellite IRNSS-1B was placed in geosynchronous orbit on 4 April 2014 aboard the PSLV-C24 rocket from Satish Dhawan Space Centre, Sriharikota. The satellite will provide navigation, tracking and mapping services.

The IRNSS-1B satellite weighs 1,432 kg and has two payloads: a navigation payload and CDMA ranging payload in addition with a laser retro-reflector. The payload generates navigation signals at L5 and S-band. The design of the payload makes the IRNSS system interoperable and compatible with GPS and Galileo. The satellite is powered by two solar arrays, which generate power up to 1,660 watts, and has a life-time of ten years.

### IRNSS-1C

Satellite IRNSS-1C was placed in geostationary orbit on 16 October 2014 aboard PSLV-C26 from the Satish Dhawan Space Centre, Sriharikota.

The IRNSS-1C satellite has two payloads: a navigation payload and CDMA ranging payload in addition with a laser retro-reflector. The payload generates navigation signals at L5 and S-band. The design of the payload makes the IRNSS system interoperable and compatible with GPS and Galileo. The satellite is powered by two solar arrays, which generate power up to 1,660 watts, and has a life-time of ten years.

### IRNSS-1D

IRNSS-1D is the fourth IRNSS satellite. It was launched using India's PSLV-C27 on 28 March 2015.

### IRNSS-1E

IRNSS-1E is the fifth IRNSS satellite. It was launched on 20 January 2016 using India's PSLV-C31.

### IRNSS-1F

IRNSS-1F is the sixth IRNSS satellite. It was launched on 10 March 2016 using India's PSLV-C32.

### IRNSS-1G

IRNSS-1G is the seventh IRNSS satellite. It was launched on 28 April 2016 using India's PSLV-C33, which concludes the setting up of the Indian Regional Navigation Satellite System.

## Quasi-zenith Satellite System

Quasi-Zenith satellite orbit

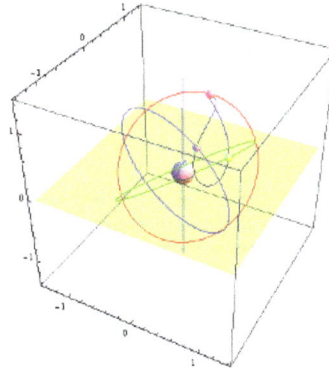

QZSS animation

The Quasi-Zenith Satellite System (QZSS), is a proposed three-satellite regional time transfer system and the satellite-based augmentation system for the Global Positioning System, that would be receivable within Japan. The first satellite 'Michibiki' was launched on 11 September 2010. Full operational status was expected by 2013. In March 2013, Japan's Cabinet Office announced the expansion of the Quasi-Zenith Satellite System from three satellites to four. The $526 million contract with Mitsubishi Electric for the construction of three satellites is slated for launch before the end of 2017. The basic four-satellite system is planned to be operational in 2018.

Authorized by the Japanese government in 2002, work on a concept for a Quasi-Zenith Satellite System (QZSS), or *Juntencho* (準天頂?) in Japanese, began development by the Advanced Space Business Corporation (ASBC) team, including Mitsubishi Electric, Hitachi, and GNSS Technologies Inc. However, ASBC collapsed in 2007. The work was taken over by the Satellite Positioning Research and Application Center. SPAC is owned by four departments of the Japanese government: the Ministry of Education, Culture, Sports, Science and Technology, the Ministry of Internal Affairs and Communications, the Ministry of Economy, Trade and Industry, and the Ministry of Land, Infrastructure and Transport.

QZSS is targeted at mobile applications, to provide communications-based services (video, audio, and data) and positioning information. With regards to its positioning service, QZSS can only provide limited accuracy on its own and is not currently required in its specifications to work in a stand-alone mode. As such, it is viewed as a GNSS Augmentation service. Its positioning service could also collaborate with the geostationary satellites in Japan's Multi-Functional Transport Satellite (MTSAT), currently under development, which itself is a Satellite Based Augmentation System similar to the U.S. Federal Aviation Administration's Wide Area Augmentation System (WAAS).

## Orbit

QZSS uses three satellites, each 120° apart, in highly inclined, slightly elliptical, geosynchronous orbits. Because of this inclination, they are not geostationary; they do not remain in the same place in the sky. Instead, their ground traces are asymmetrical figure-8 patterns (analemmas), designed to ensure that one is almost directly overhead (elevation 60° or more) over Japan at all times.

The nominal orbital elements are:

| QZSS satellite Keplerian elements (nominal) | |
|---|---|
| Epoch | 2009-12-26 12:00 UTC |
| Semimajor axis ($a$) | 42,164 km |
| Eccentricity ($e$) | 0.075 ± 0.015 |
| Inclination ($i$) | 43° ± 4° |
| Right ascension of the ascending node ($\Omega$) | 195° (initial) |
| Argument of perigee ($\omega$) | 270° ± 2° |
| Mean anomaly ($M_o$) | 305° (initial) |
| Central longitude of ground trace | 135° E ± 5° |

## QZSS and Positioning Augmentation

The primary purpose of QZSS is to increase the availability of GPS in Japan's numerous urban canyons, where only satellites at very high elevation can be seen. A secondary function is performance enhancement, increasing the accuracy and reliability of GPS derived navigation solutions.

The Quasi-Zenith Satellites transmit signals compatible with the GPS L1C/A signal, as well as the modernized GPS L1C, L2C signal and L5 signals. This minimizes changes to existing GPS receivers.

Compared to standalone GPS, the combined system GPS plus QZSS delivers improved positioning performance via ranging correction data provided through the transmission of submeter-class performance enhancement signals L1-SAIF and LEX from QZSS. It also improves reliability by means of failure monitoring and system health data notifications. QZSS also provides other support data to users to improve GPS satellite acquisition.

According to its original plan, QZSS was to carry two types of space-borne atomic clocks; a hydrogen maser and a rubidium (Rb) atomic clock. The development of a passive hydrogen maser for QZSS was abandoned in 2006. The positioning signal will be generated by a Rb clock and an architecture similar to the GPS timekeeping system will be employed. QZSS will also be able to use a Two-Way Satellite Time and Frequency Transfer (TWSTFT) scheme, which will be employed to gain some fundamental knowledge of satellite atomic standard behavior in space as well as for other research purposes.

## QZSS Timekeeping and Remote Synchronization

Although the first generation QZSS timekeeping system (TKS) will be based on the Rb clock, the first QZSS satellites will carry a basic prototype of an experimental crystal clock synchronization system. During the first half of the two year in-orbit test phase, preliminary tests will investigate the feasibility of the atomic clock-less technology which might be employed in the second generation QZSS.

The mentioned QZSS TKS technology is a novel satellite timekeeping system which does not require on-board atomic clocks as used by existing navigation satellite systems such as GPS, GLONASS or the planned Galileo system. This concept is differentiated by the employment of a synchronization framework combined with lightweight steerable on-board clocks which act as transponders re-broadcasting the precise time remotely provided by the time synchronization network located on the ground. This allows the system to operate optimally when satellites are in direct contact with the ground station, making it suitable for a system like the Japanese QZSS. Low satellite mass and low satellite manufacturing and launch cost are significant advantages of this system. An outline of this concept as well as two possible implementations of the time synchronization network for QZSS were studied and published in and.

# DORIS (Geodesy)

Doppler Orbitography and Radiopositioning Integrated by Satellite or, in French, Détermination d'Orbite et Radiopositionnement Intégré par Satellite (in both case yielding the acronym DORIS) is a French satellite system used for the determination of satellite orbits (e.g. TOPEX/Poseidon) and for positioning.

## Principle

Ground-based radio beacons emit a signal which is picked up by receiving satellites. This is in reverse configuration to other GNSS, in which the transmitters are space-borne and receivers are in majority near the surface of the Earth. A frequency shift of the signal occurs that is caused by the movement of the satellite (Doppler effect). From this observation satellite orbits, ground positions, as well as other parameters can be derived.

## Organization

DORIS is a French system which was initiated and is maintained by the French Space Agency (CNES). It is operated from Toulouse.

## Ground Segment

The ground segment includes about 50-60 ground stations, equally distributed over the earth and ensure a good coverage for orbit determination. For the installation of a beacon only electricity is required because the station only emits a signal but does not receive any information. DORIS beacons transmit to the satellites on two UHF frequencies, 401.25 MHz and 2036.25 MHz.

The UHF transmitting antenna of DORIS ground station at Dionysos, Greece

## Space Segment

The best known satellites equipped with DORIS receivers are the altimetry satellites TOPEX/Poseidon, Jason 1 and Jason 2. They are used to observe the ocean surface as well as currents or wave heights. DORIS contributes to their orbit accuracy of about 2 cm.

Other DORIS satellites are the Envisat, SPOT, HY-2A and CryoSat-2 satellites.

## Positioning

Apart from orbit determination, the DORIS observations are used for positioning of ground stations. The accuracy is a bit lower than with GPS, but it still contributes to the International Terrestrial Reference Frame (ITRF).

## References

- "IRNSS-1G exemplifies 'Make in India', says PM". The Statesman. 28 April 2016. Retrieved 28 April 2016.

- "India successfully launches IRNSS-1D, fourth of seven navigation satellites - Times of India". indiatimes.com. Retrieved 7 September 2016.

- "India launches 5th navigation satellite IRNSS-1E powered by PSLV rocket". hindustantimes.com. Retrieved 2016-01-20.

- Narasimhan, T. E. "India gets its own GPS with successful launch of 7th navigation satellite". business-standard.com. Retrieved 7 September 2016.

- K. Radhakrishnan (29 December 2013). "Mars and more, final frontier". Deccan Chronicle. Retrieved 28 April 2016.

- "'All seven satellites of IRNSS to be in orbit by March 2016'". Business Standard. 29 October 2015. Retrieved 1 November 2015.

- "'India launches seventh navigation satellite, to get its own GPS'". International Business Times. 28 April 2016. Retrieved 28 April 2016.

- Srivastava, Ishan (5 April 2014). "How Kargil spurred India to design own GPS". The Times of India. Retrieved 9 December 2014.

- "ISRO: After GSLV launch, PSLV C24 with IRNSS-1B likely in March". India TV News. 5 January 2014. Retrieved 6 January 2014.

- "IRNSS". space.skyrocket.de. Retrieved 6 December 2014. Cite error: Invalid <ref> tag; name "skyrocket" defined multiple times with different content.

- Fabrizio Tappero (April 2008), Remote Synchronization Method for the Quasi-Zenith Satellite System (PhD thesis), retrieved 2013-08-10

- "ISRO opens navigation centre for satellite system". Zeenews.com. 2013-05-28. Retrieved 30 June 2013.

# Permissions

All chapters in this book are published with permission under the Creative Commons Attribution Share Alike License or equivalent. Every chapter published in this book has been scrutinized by our experts. Their significance has been extensively debated. The topics covered herein carry significant information for a comprehensive understanding. They may even be implemented as practical applications or may be referred to as a beginning point for further studies.

We would like to thank the editorial team for lending their expertise to make the book truly unique. They have played a crucial role in the development of this book. Without their invaluable contributions this book wouldn't have been possible. They have made vital efforts to compile up to date information on the varied aspects of this subject to make this book a valuable addition to the collection of many professionals and students.

This book was conceptualized with the vision of imparting up-to-date and integrated information in this field. To ensure the same, a matchless editorial board was set up. Every individual on the board went through rigorous rounds of assessment to prove their worth. After which they invested a large part of their time researching and compiling the most relevant data for our readers.

The editorial board has been involved in producing this book since its inception. They have spent rigorous hours researching and exploring the diverse topics which have resulted in the successful publishing of this book. They have passed on their knowledge of decades through this book. To expedite this challenging task, the publisher supported the team at every step. A small team of assistant editors was also appointed to further simplify the editing procedure and attain best results for the readers.

Apart from the editorial board, the designing team has also invested a significant amount of their time in understanding the subject and creating the most relevant covers. They scrutinized every image to scout for the most suitable representation of the subject and create an appropriate cover for the book.

The publishing team has been an ardent support to the editorial, designing and production team. Their endless efforts to recruit the best for this project, has resulted in the accomplishment of this book. They are a veteran in the field of academics and their pool of knowledge is as vast as their experience in printing. Their expertise and guidance has proved useful at every step. Their uncompromising quality standards have made this book an exceptional effort. Their encouragement from time to time has been an inspiration for everyone.

The publisher and the editorial board hope that this book will prove to be a valuable piece of knowledge for students, practitioners and scholars across the globe.

# Index